# NON-FORMAL EDUCATION FOR INTEGRATED PRODUCTION AND PEST MANAGEMENT TRAINING

# NON- FORMAL EDUCATION FOR TRAINING IN INTEGRATED PRODUCTION AND PEST MANAGEMENT IN FARMER FIELD SCHOOLS

**ALBERT D. K. AMEDZRO**
**ANTHONY YOUDEOWEI**

GHANA UNIVERSITIES PRESS
ACCRA
2005

*Published by*
**Ghana Universities Press**
*(Universities Publishing Foundation)*
P. O. Box GP 4219
Accra
Tel. 233 (021) 513401, 513383, 513404
Fax. 233(021) 513402
E-mail: ghanauniversitiespress@yahoo.com

© Albert D. K. Amedzro and Anthony Youdeowei, 2005
ISBN: 9964-3-0314-9

**PRODUCED IN GHANA**
Typesetting by Ghana Universities Press, Accra, Ghana.
Printing and binding by Raspa Press Ltd Accra
Ghana   Tel: 021 - 221772

# CONTENTS

*Preface* ix
*Acknowledgement* xi

**Chapter 1** 1
**Non- Formal Education in Integrated Production and Pest Management (IPPM) Training Programmes** 1

    The objectives of this chapter 1
    Introduction 1
    The Objective of this Field Guide 3
    Using this Field Guide and Tips for Conducting Non-Formal
        Education Sessions 3
    Techniques for Team Building and Participation 5

**Chapter 2** 7
**Education and Training in IPPM** 7
    Objectives of this chapter 7
    Differences between Education and Training 7
    Importance of Training 8
    Characteristics of Group Training 12
    The Environment for Sharing Knowledge amongst
        Farmers 12

**Chapter 3** 20
**General Principles of Adult Learning and Teaching** 20
    The objectives of this chapter 20
    Profile of Adult Learners 20
    Characteristics of Adult Learners 21
    Principles of Adult Learning and Teaching 23
    People's Values 30

**Chapter 4** 34
**Approaches and Methods of Non-Formal Education** 34
    The objectives of this chapter 34
    Forms of Education 34

| | |
|---|---|
| Advantages of Non-Formal Education | 34 |
| Non-Formal Education Techniques and Methods | 36 |
| Qualities of a Good Facilitator | 41 |

## Chapter 5
## The Communication Process

| | |
|---|---|
| | 42 |
| | 42 |
| The objectives of this chapter | 42 |
| The Importance of Communication in the Facilitation Process | 42 |
| Interpersonal Communication | 43 |
| Directional Ways of Communication | 44 |
| Forms of Communication | 45 |
| Channels of Communication | 48 |
| Types of Communication | 48 |
| Routes of Communication | 49 |
| Barriers to Effective Communication | 50 |
| Enhancing the Communication Process | 52 |

## Chapter 6
## Interpersonal Relationship: Johari's Window

| | |
|---|---|
| | 54 |
| | 54 |
| The objectives of this chapter | 54 |
| Introduction | 54 |
| The Free/Open Window | 54 |
| The Blind Window | 56 |
| The Hidden Window | 58 |
| The Dark Window | 61 |

## Chapter 7
## Effective Teamwork

| | |
|---|---|
| | 63 |
| | 63 |
| The objectives of this chapter | 63 |
| A Team | 63 |
| Effective Teamwork | 63 |
| Aims and Objectives of Teams | 64 |
| Decision Making Procedures | 64 |
| Relationships within the Team | 65 |
| Reviewing Team Performance | 68 |
| Relationships with Outsiders | 68 |

**Chapter 8** 69
**Leadership** 69
    The objectives of this chapter 69
    Introduction 69
    Leadership Styles 70
    Four Leadership Styles 71
    Human Behaviour within Groups 78

**Chapter 9** 94
**Decision Making in Integrated Production and Pest Management: Agro-Ecosystem Analysis** 94
    The objectives of this chapter 94
    Introduction 94
    Exercises 95

**Chapter 10** 101
**Brighten Your Training Sessions** 101
    The objectives of this chapter 101
    Introduction 101
    Jokes 101
    Proverbs 107
    Some Thoughts for the Day 110
    Reflections 111
    Wise Sayings Associated with Great Men 113
    Ten Important Guides to Fulfilling Life 115

*Recommended Reading* 117

# PREFACE

The aim of this Field Guide is to provide simple exercises for IPPM Trainers and Facilitators who are actively involved in season-long, farmer-based training in the implementation of Integrated Production and Pest Management methodologies.

In preparing the guide, we have drawn on the vast experiences from IPPM master trainers and materials from the National IPPM programmes of Asian countries where the FAO Inter-Country IPPM Programme developed the IPPM Farmers' Field School concept. The Asian experience has been appropriately modified and applied to local field and farming community situations in Ghana. Thus, the guide has benefited from the experiences of conducting training and implementation of IPPM in Ghana.

Users of the guide will, therefore, find that the examples given are specific to the Ghanaian farming environment. However, since the field conditions in Ghana are, to a large extent, similar to those of other West African countries, we hope that the guide will assist in the season-long training and implementation of IPPM in other countries of the West African sub-region. The guide can therefore be adopted for use by all categories of facilitators and trainers.

# ACKNOWLEDGEMENTS

This Field Guide was developed during various training events in Integrated Production and Pest Management in Farmer Field Schools organized for agricultural extension agents and small holder farmers in Ghana. The programme was supported by the UNDP/ FAO under the National Poverty Reduction, Agro- skills development project for farmers.

Several examples and exercises have been taken directly or modified from excellent materials published in *Training for Transformation* by Ann Hope and Sally Timmel (Mambo Press), and *Group Process and the Inductive Method* by Carmela D. Ortigas. We are grateful for their kind permission to use these materials in this Field Guide.

We are grateful to the UNDP, the FAO and the Global IPPM Facility based at FAO Headquarters in Rome, for the opportunity to participate in this most exciting programme. Our very special thanks are due to the agricultural extension agents who were trained as IPPM Trainers/ Facilitators and the numerous small holder farmers who allowed us to intrude into thier farming environment and accommodated the pressure of our training activities. The success achieved during the training programme was largely due to the excellent cooperation from the farmers who have now become integrated production and pest management experts in their own fields. We hasten to absolve them from any shortcomings in this guide for which we take full responsibility.

It is our hope that this Guide will be helpful to IPPM Trainers and Facilitators who are involved in farmer training efforts to improve the lives and well- being of African farmers.

CHAPTER 1

# Non-formal Education in Integrated Production and Pest Management (IPPM) Training Programmes

**The objectives of this chapter are:**

1. to provide a background to the role of non-formal education in IPPM training.
2. to explain the purpose and use of this field guide.

**Introduction**

> I have the audacity to believe that peoples everywhere can have three meals a day for their bodies, education and culture for their minds, and dignity, equality, and freedom for their spirits.
>
> The Rev. Martin Luther King

These are the famous words of the late Reverend Martin Luther King in his Nobel Prize Acceptance Speech. He believed that:

1. without relevant education, many people would be denied the basic comforts of life, and they will not obtain and enjoy a balanced diet;

2. without adequate education, people will be culturally tamed; they will forfeit their freedom and their spirits will be dampened;

3. the only salvation for any people to keep their bodies and souls nourished and brought together is to participate in some form of education.

All persons who are involved in community work, as in Integrated Production and Pest Management (IPPM) training in Farmer Field Schools, are always confronted with problems which relate to the acquisition of human needs including food, shelter, and the ability to make personal decisions and to control their own lives. Through understanding people's needs, the way they feel, and their learning process, it is possible and easier to educate them and to introduce innovations that will lead to visible improvements in their lives. In community development language, we say that education and development liberate people.

One of the major objectives of IPPM training in Farmer Field Schools is to empower farmers to make crop and pest management decisions in their own fields and to take appropriate corrective actions where necessary. As farmers become empowered, they gradually begin to take responsibility for shaping their environment, their lives, their community and society. This kind of action can be appropriately described as a form of liberation. IPPM-trained farmers are "liberated" from blindly adopting standard technology packages that have been developed usually in the absence of, and often without involving the farmers who are the target beneficiaries of the technologies. Empowered farmers develop the capacity for careful ecological and scientific observations of their crop production ecosystem and critical analysis of the factors which influence crop growth and production in order to make management decisions based on the results of observation and analysis. With such analytical minds, farmers begin to introduce new concepts to improve their societies and to build new communities.

According to Hope and Timmel (1984), the skills required to build new communities and societies, include the need to:

1. improve our communication,
2. learn to listen,
3. express our insights,
4. diagnose together our needs,
5. analyze the causes of our problems and
6. plan and act together in teams, organizations and movements.

Farmers need these skills in order to effectively contribute to the betterment of their communities and to drastically raise their standards of living. The development of these skills is a primary process in IPPM training in Farmer Field Schools. Non-formal education techniques are designed to assist community workers to facilitate the development of these skills, in this case, by smallholder farmers. Non-formal education, therefore, becomes a necessary and essential component of the curriculum of IPPM Training of Trainers and Farmers in Farmer Field Schools.

## The Objective of this Field Guide

This field guide is designed to introduce the IPPM trainer or facilitator to key aspects of non-formal education that will assist in understanding the ways in which adults learn and how sustainable farmer groups can be formed. The guide contains basic ideas, which should be modified according to the local situations where the training is conducted.

## Using this Field Guide and Tips for Conducting Non-formal Education Sessions

IPPM trainers and facilitators should be innovative and adapt the materials here as much as possible, to local conditions, languages, and cultures. They need to:

1. *Conduct practical exercises*
   IPPM trainers and facilitators who use this Field Guide should appreciate that practical exercises are very important for reinforcing the theoretical aspects of any of the non-formal education topics. They should, therefore, make considerable effort to conduct the exercises immediately the theoretical aspects of a topic are treated.

2. *Adapt exercises to be culture-relevant*
   The practical exercises should relate particularly to current issues in the locality and to the cultural norms of the farming community where training is conducted. This helps to bring into sharp focus the concepts of adult learning, teamwork and leadership, and inter-personal relationships and group/community action.

3. *Create a friendly environment*
   Adults learn best under an informal, friendly, and partnership environment. Trainers and facilitators need, therefore, to make the special effort to create a friendly atmosphere, which is conducive for learning. The exercises suggested in this Field

Guide provide ample opportunities for fun and games. Additional exercises and icebreakers should be introduced frequently during the training sessions.

4. *Adopt a fully-participatory approach*
   Encourage full participation by all members of the group so that everyone feels an integral part of the action and the group. Nobody is the "Expert " or the "Smart Alec" who has all the answers or the only bright ideas! When you think critically about knowledge, you will realize that nobody, especially farmers, is really totally ignorant. Like the broken-down clock, which is correct twice in the day, once in the morning and then in the afternoon, no one person is completely useless.

   This concept recognizes the value of the wisdom of farmers, enables them to gain self-confidence and engenders community spirit.

5. *Set time limits*
   Prepare a time frame for the entire programme. Set a time limit for each of the topics and try to work within it. Try to complete the topic and reach a decision within the time allocated to each period. You can be flexible if interest in a particular topic is high.

6. *Follow an agenda*
   Introduce one topic at a time and involve farmers fully in all discussions. Always direct the discussion to focus on the particular topic. Do not allow participants to deviate too much although they may express opinions that may sometimes be too personal and irrelevant to the topic being discussed.

7. *Ask many questions*
   One good way of involving participants in non-formal education programmes is to ask questions. This method is known as the *Socratic technique*. It derives its name from an ancient technique of learning used by the Greek philosopher, Socrates. The objective of the Socratic method is to reduce the boredom of lecturing, which at times degenerates into talking down to adults.

The method engages adult learners actively in discussing and discovering for themselves truths and principles. The trainer makes challenging and controversial statements to which learners respond. The probing questions continue further until a consensus is reached or solutions are found to problems.

The Socratic technique requires the facilitator to carefully plan the pattern of questioning and discussion, drawing on the experiences from the immediate environment. The questions draw out people's ideas on the problem but do not provide direct suggestions or solutions.

8. *Clarify issues*
   There may be doubts about some contributions that people make. Some statements may not be clear. Do not assume that the majority of participants follow the trend of a discussion accurately. Ask questions or repeat a point to clarify issues; you may often provide examples to illustrate suggestions.

9. *Summarize discussion and consensus*
   When a topic is fully discussed, the trainer or facilitator should summarize the main points of discussion and consensus, emphasizing the main issues raised and the various lines of action prescribed.

10. *Plan for action*
    The trainer or facilitator assists the group to adopt strategies to implement the agreed plan. Members are assigned roles on **what** to do, **who** does it, **why**, **when**, **how** it should be done, which materials to use and where to do things.

## Techniques for Team Building and Participation

Four examples of techniques for team building and participation are:

1. Use different techniques in your facilitation work.

2. Use different group sizes to encourage active participation of

all learners. Small working groups can be composed of five to eight learners, or buzz groups of pairs. The composition of the groups should be changed often in order to achieve full interaction among participants.

3. Encourage everybody to speak freely, praising each contribution with words such as, "Good", "Well done", " Excellent", "That is a brilliant idea". You may also nod your head in appreciation. If an answer is wrong, the trainer should not tell the presenter directly that he/she is wrong; he should be diplomatic by asking if anybody else has an alternative suggestion. In this way, the presenter will realize that the answer originally provided was not quite correct.

4. Adopt a friendly look, relax, smile and crack moderate jokes to attract participation. Some sample jokes, proverbs and thoughts for the day are included in Chapter 10 of this Field Guide to assist trainers make their sessions lively and interesting. As much as possible, trainers should avoid talking too much.

CHAPTER 2

# Education and Training in IPPM

The objectives of this chapter are:
1. to explain the basic differences between education and training.
2. to understand the characteristics of groups and knowledge sharing among group members.

## Differences between Education and Training

When working with adults, especially farmers, it is necessary for IPPM trainers and facilitators to understand clearly the differences between education and training and how these differences relate to the farmer's situation.

The term **training** describes a process of providing knowledge and skills in order to bring about desired changes in people's attitudes and to improve their competencies. The primary goal of training is, therefore, to improve performance. Education also provides knowledge and brings about changes in attitudes and skills. However, education differs from training because education is aimed at preparing people to meet the challenges of the future.

Some major differences between training and education are summarized as in Table 2.1.

TABLE 2.1

**Major Differences between Education and Training**

| *Education* | *Training* |
| --- | --- |
| 1. Education is usually long-term | Training is usually short-term |
| 2. Education is broadly focused | Training is specific and narrowly focused |
| 3. Education is usually aimed at preparing people to cope with the challenges of the future | Training is usually designed to meet specific needs and has the potential of immediate application |

An important aspect of training is the Training of Trainers' Programme. Since it is difficult to train all members of a target community of a programme at the same time, the ideal situation is to train the leaders who constitute a core of trainers to train others. This strategy for training has its merits. It is cost effective and content intensive. More importantly, its multiplier effects are great as the training graduates from one level to the other.

Note that some authors use the term education to refer to training. Training, like education, has many advantages as pointed out by some authors including Beardwell and Holden (1994).

**Importance of Training**

Training,

1. serves as an incentive to workers;
2. attracts and maintains good workers in an organization;
3. equips employees with appropriate skills within a short time;
4. sharpens the skills of existing employees and keeps them abreast with modern methods of production and new managerial procedures;
5. reduces obsolesce of workers;
6. reduces the cost of employment of new personnel;
7. reduces accidents, mistakes and wastage;
8. leads to quality and quantity improvement in goods;
9. prepares workers to be more receptive to changes (new machines, technology and methods);
10. gives job and economic satisfaction to employees;
11. gives relief to management to have peace of mind to concentrate on other development issues;
12. boosts the morale of workers;
13. builds cordial relationships between trainees and trainers; and
14. enhances workers' prospects for promotion.

How do these terms relate to IPPM training in Farmer Field Schools situation? According to Freire (1970), no education is neutral; education is either designed to domesticate people, that is, to maintain the *status quo* when a dominant class imposes its will, values and culture on the people as for example during imperialist colonization, or to liberate them (*see* Fig. 2.1). When education programmes are designed to liberate people,

*Fig. 2.1:* Education is either a domesticating (left) or a liberating experience (right). (adapted from Hope and Timmel, 1984)

beneficiaries become more critical, ask many questions, are proactive, free, creative and take full responsibility for shaping their lives and environment. Another important aspect of education is that people react more positively to issues that are of relevance to them, so whatever people learn should be of immediate relevance and concern to them.

In IPPM training, the issues of immediate relevance to farmers are high crop yields and profitable agricultural production in a healthy environment. The aim of Integrated Production and Pest Management Training in Farmer Field Schools is, therefore, to liberate farmers so that they become knowledgeable about their farming environment and can confidently take charge of shaping their farming activities and lives in their own communities.

In IPPM training in Farmer Field Schools, trainers and facilitators are involved in training of farmers, based on a season-long specific training curriculum to meet a specific need and to educate farmers in order to empower them with knowledge and skills to improve their future crop production efforts. Training farmers to conduct agro-ecosystem analysis is an exercise in developing their capacity to participate in a common search for solutions to identified problems. All the farmers are recognized as thinking and creative people, who have some knowledge to share with other members of their groups. Training farmers to conduct agro-ecosystem analysis and other exercises enable them to understand the principles of doing things with people rather than the trainer or facilitator doing things for them. By doing things themselves, farmers become less dependent on others and develop the spirit of self-reliance.

## Exercise 2.1 Doing things with people

*Purpose:*
This exercise is conducted through a role-play in which participants are presented with a situation, which helps them to realize that *doing things with people* is more successful and rewarding than *doing things for people*.

*Materials Required:*
Chalk, pieces of paper, a large piece of newsprint

*Procedure:*
1. Draw two parallel lines on the floor one meter apart. These represent the banks of a deep river. Draw shorter wavy lines in between the long lines to represent flowing water (*see* Fig. 2.2). Place the small pieces of paper on the river to represent stepping stones that should aid in crossing the river. Place the large newsprint to represent an island in the middle of the river.

*Fig. 2.2:* Doing things with people is more beneficial than doing things for people. (adapted from Hope and Timmel, 1984)

2. Invite three participants and explain to them the situation presented in Box 2.1. They should then perform this role-play without speaking a word (mime), while the rest of the participants watch.

> **BOX 2.1**
>
> **Description of Exercise 2.1 Role Play**
>
> Two men approach the bank of the river and look for a place to cross. But seeing the fast current and the width of the river, they are afraid to cross. A third man, the ferryman, comes along, leads them up to the bank of the river and shows them the stepping stones and encourages them to step on the stones to cross the river.
>
> They are still afraid and refuse to cross. He then agrees to carry one of the men on his back to cross the river, but when he gets to the middle of the river the man on his back is so heavy that he cannot continue to carry him. He therefore decides to put him down on the island in the middle of the river. The man is left on the island, shivering and showing signs of desperation. The ferryman explains that he can no longer continue to carry him because the burden of his weight is too much for him to cope with.
>
> The ferryman now returns to fetch the other man who is still standing by the bank of the river waiting to be carried to cross the river. The ferryman refuses to carry him, instead he holds his hand and encourages him to follow as he steps on the stones to cross the river. Working with each other, both successfully cross the river and on getting to the other side, they are very pleased with themselves and walk off together leaving the other man sitting frustrated on the island in the middle of the river. He tries very hard to attract their attention but they do not take any notice of his cry for help.

At the end of this role-play, the players should return to take their seats. The facilitator should conduct an open discussion of what happened. The discussion should start with the description of what happened, in the play, before the analysis of what happened. The discussion ends with a conclusion showing the lessons learned.

The facilitator should not allow any one person to dominate the discussion. For example, in the first question, one person should not be allowed to describe all the scenes. As the discussion of the role-play proceeds, one participant is asked to write the main lessons learned from the play on the flip chart. These are later pasted on the wall of the classroom.

*Discussion Questions:*
1. Describe what happened in this role-play.
2. What different approaches were adopted to help the two men cross the river?
3. Why was the man who was abandoned on the island in the middle of the river frustrated?
4. What is the main principle in this role-play?
5. What happens when we *work for people*?
6. What happens when we *work with people*?
7. How often do similar situations happen in real life?
8. What does this role play tell us about education and training to liberate people?

## Characteristics of Group Training

Training people in groups has considerable advantages over individual training. The main advantages of group training are:

1. Group training encourages learning through sharing of knowledge and experiences amongst trainees.

2. In effective group training, all participants are given ample chance to speak their own words and to contribute to the training session.

3. Because people have different perceptions and experiences, training them in groups enriches their perceptions, knowledge and skills.

## The Environment for Sharing Knowledge amongst Farmers

Training sessions in Farmer Field Schools should be conducted in a very relaxed and friendly atmosphere. Research on adult learning shows that such a training environment promotes effective learning. Under such conditions, farmers share their knowledge and experiences more readily; they listen to and learn from their colleagues. Trainers and facilitators, therefore, need to create a conducive learning environment which will help farmers to express themselves freely while at the same time learn from other farmers. Through this process, farmers discover for themselves rather than being informed by the facilitator.

Research has also shown that people remember 20 per cent of what they Hear, 40 per cent of what they See and Hear and 80 per cent of what they Discover for themselves (Hope and Timmel, 1984). *So the learning by discovery mode is a very important pattern to be adopted for IPPM training in Farmer Field Schools.*

To successfully share information and knowledge, adults in-groups need to build a high level of trust amongst themselves. Gibbs identified four needs of a group. These are:

1. Acceptance
2. Sharing information and concerns
3. Goal setting
4. Organizing for group action.

If there is a spirit of mutual respect and acceptance, people become free to learn and to fully share their knowledge and experiences. Group members need information about one another, their perceptions, ideas, experiences and values, and about the issues that they consider most important in their lives. Goals set by members of a group are accepted by those who tend to be committed to achieving such goals. And once goals are collectively set, the group usually makes plans to achieve these goals. Facilitators should, therefore, ensure that these group needs are addressed in order to create a conducive climate for farmers to share knowledge, learn and work together.

### *Exercise 2.2 Knowing each other*

*Purpose:*
An essential pre-requisite for establishing a friendly atmosphere amongst a group of people is for them to know each other. The purpose of this exercise is, therefore, to establish rapport among farmers and to develop a friendly training environment.

*Materials Required:*
Drawings or magazine pictures of farming or community objects, each cut into two halves.

Examples are pictures of a cow, a donkey, a cabbage plant, a tomato plant, a rice plant, spiders, caterpillars, beetles, a village house, and a lorry.

Envelopes, Sheets of newsprint, coloured markers or crayons.

*Procedure:*
Place each half of the cut pictures in separate envelopes.

Spread the envelopes on a table.

Brief the participants as follows:

Pick one envelope; look at the picture in the envelope and locate the person with the other half of the picture.

When you find the person with the other half of your picture, sit together in one corner of the room and paint the picture as you wish.

Get to know yourselves by asking questions such as:

Give me some information about yourself; you know that we are going to work together during this training so I think that we should get to know ourselves better. Don't you think so? I will tell you many things about myself too. Some of the questions to raise are:

1. What is your name?
2. From which village or community do you come?
3. What do you do for a living in addition to farming?
4. What crops do you grow and what animals do you keep?
5. How many wives and children do you have?
6. What do you like about your community?
7. Is this your first training?
8. If No, in which other training have you been involved?
9. What hobby do you do best?
10. What food do you like best?

Give about 15 minutes for participants to interview each other.
All participants should now come together and sit in a circle so that they

can all see themselves easily to allow easy interaction.

Ask each person to introduce the new-found friend to the rest of the group.

Each person should spend about three minutes for this introduction.

*Discussion Questions:*
1. Are you happy that you have met a new friend?

2. Do you think that you have obtained a lot of information about your new friend?

3. Did you give interesting information about yourself to your new friend?

4. Do you have many things in common with your new friend?

5. Now that you know your new friend, do you believe that you will be comfortable working together?

### *Exercise 2.3 Sharing knowledge and information amongst farmers*

*Purpose:*
Participants now know one another and have initiated some form of friendship and trust amongst themselves. The purpose of this exercise is to understand how to create a friendly and favourable climate for sharing knowledge and information among farmers.

*Materials Required:*
Newsprint, coloured markers or crayons, plywood boards to stick the newsprint and cellotape.

*Procedure:*
1. Divide the group of farmers into small working groups of about 5 persons each (*see* Fig. 2.3).

2. Hold a short briefing separately with each group and tell them to discuss any topic, which they consider very important to their lives, work, or community.

3. Allow 20 minutes for the small group discussions.

4. After 20 minutes, stop the small group discussion and ask participants to re-assemble as one large group.

5. Through a reporter, each group should now present what was discussed to the larger group.

6. The facilitator should write down on newsprint, the main issues discussed by each group.

*Fig. 2.3:* The group of farmers is divided into small working groups.

*Discussion Questions:*
These questions are meant to find out how well members of each group understood the content of the discussion. If a friendly environment was created, all the farmers in each group would be completely familiar with the content of the discussion in their groups.

1. Who led the discussion in the group?
2. How was the leader selected?
3. How many subjects were discussed?
4. What was the main subject discussed?
5. What conclusions were reached at the end of the discussion?
6. Were all members of the group friendly to one another during the discussion?
7. Did individual members of the group learn new things from the other members?
8. What factors promoted the free flow of information amongst the farmers?

**Exercise 2.4 How would you like to feel in your groups in the Farmer Field School?**

*Purpose:*
This exercise is a simple drawing puzzle which is designed to explore the ways farmers want to feel in their groups during the Farmer Field School (FFS) sessions.

*Materials Required:*
Sheets of paper, Large Newsprint, Coloured markers or crayons, Drawing of puzzle as illustrated here (*see* Fig. 2.4).

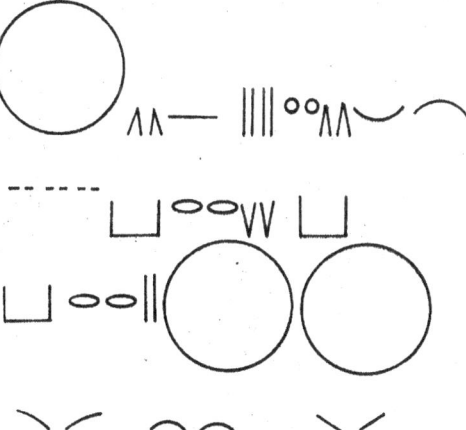

***Fig. 2.4:*** *Elements of the picture puzzle.*

*Procedure:*
1. Participants should sit in their small working groups.

2. Supply each group with three sheets of paper and coloured markers or crayons.

3. Draw the elements of the puzzle on the large newsprint and paste the drawing in front of the groups.

4. Each group should now compose pictures to show how they would like to feel during the FFS sessions. They should compose three pictures using the elements. Allow about 25 minutes for this exercise.

5. When they complete the exercise, one member of each group should make a presentation of what they have drawn.

6. Hold a large group discussion of the results of this exercise and write the list of how participants like to feel at the FFS sessions.

7. Paste the participants' pictures on the wall of the classroom throughout the FFS training.

8. Fig. 2.5 shows suggestions of answers to this puzzle.

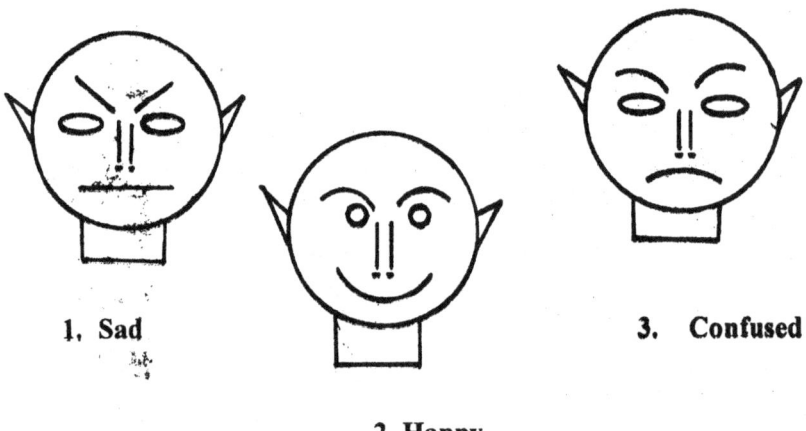

1. Sad    2. Happy    3. Confused

*Fig. 2.5:* Suggested answers to the picture puzzle.

*Discussion Questions:*
1. How many pictures did participants draw?
2. What do the pictures represent?
3. What feelings do the pictures show?
4. Which of these feelings would they prefer during the FFS?

CHAPTER 3

# General Principles of Adult Learning and Teaching

**The objectives of this chapter are:**

1. to explain adult learning and teaching principles and
2. to highlight the importance of these principles in the adult training process.

**Profile of Adult Learners**

Adult learners constitute a hetereogeneous group. They tend to have complex personalities and a wide variety of backgrounds, needs and aspirations. In order to make an impact amongst farmers, an extension agent who acts as a trainer or facilitator, needs to know the profile and characteristics of adult learners.

*Age*
Generally, the age of farmers ranges from 15 to 65 years. This means that, extension agents work with both the very young and the old. They must, therefore, be careful about how they relate with farmers. It is not advisable to group young adults with old adults in the same working teams. The youth will be uncomfortable with the adults and vice versa. If it is inevitable that both groups should work together, you should be particularly careful to avoid a situation where one group undermines the other. African traditional beliefs that govern interactions between different age groups should always be recognized and respected.

*Sex*
Both men and women engage in farming. In most African countries, more women than men work on farms. From the time of land preparation, through planting, harvesting, storage, processing and marketing, women are actively involved. The trainer or extension agent must be aware of local norms regulating the working relationships between both sexes and conduct himself/herself accordingly. Women tend not to participate fully when grouped with men because of traditional etiquette.

## Educational Level/Work Experience

There are usually differences between the educational level and experiences of farmers and also between farmers and the extension agent. The extension agent must know the different educational levels of farmers so that training can be carefully planned to facilitate learning. The level of education among farmers may influence the level of participation in programmes and adoption of new farming technologies. The trainer/facilitator should not assume that all the farmers have attained the same levels of education or awareness and therefore treat them in the same manner. Different levels of understanding may reflect the different levels of education and work experience.

## Family Status

Adults cherish their families and seek to protect their survival. Marriage and family life are therefore very important institutions in the society. Many adults are married and have children whom they are very proud of.

## Cultural/Religious Background

An extension agent, trainer or facilitator needs to know the culture or the religious background of farmers. This will help him to know how to comport himself. Cultural and religious backgrounds strongly influence people's behaviour and levels of understanding. Traditional religious practices, Islamic precepts and Christian doctrines must be studied and appropriately respected by the facilitator.

## Socio-economic Status

Farmers have various social roles to play since they are parents, spouses, citizens, community and church leaders, members and leaders of social and traditional groups, friends and neighbours. They therefore require a variety of educational needs to face the challenges of social and technological changes.

# Characteristics of Adult Learners

Adult learners have physical, physiological, psychological and sociological characteristics which affect their learning. Physiologically, as adults advance in years, they go through a series of physical changes such as loss of teeth, poor eye sight and sluggishness. Muscular control is difficult to achieve by old people. Twisting hand movements could be painful. Tiredness and pregnancy may cause tardiness.

Sociologically, adults have many community and social responsibilities. They may be landlords, grandfathers and grandmothers and be members of many community groups such as women groups and professional associations. They are fathers and mothers with parental responsibilities. They, therefore, expect to be given the respect and dignity attached to their status. Many adults drop out from agricultural extension programmes and other educational activities because of how young adult facilitators treat them. They admire the social system and structure within which they operate. Adults enter any learning situation with a different background and experience from that of the youth. Having lived longer, adults define themselves largely by their experience, they have a deep investment in it.

Psychologically, adults may have memory problems since many needs compete for their attention. Some may have the fear of failure and have aversion to competition; others may also be reminded of the painful memories of the stress of their school days. By and large, however, if given suitable conditions, an adult can go on learning and is prepared to learn throughout life.

*Farmers' Needs*

The extension agent or facilitator should note that his/her clients have a variety of needs and perceptions that influence their participation in programmes. Their prime motive may be to learn techniques that will help them improve upon performance in their farms so that they can earn a decent living. The extension agent should try to create an environment for farmers to realize some of the basic human needs discussed by Maslow (1970).

According to Maslow, cited in Knowles (1970), the basic needs of humans can be categorized as:

1. Physiological or survival needs: food, shelter, and clothing (basic needs).
2. Safety needs: (Security), Protection from physical harm; protection by the law, friends, society and co-operatives, provision of job opportunities and economic comfort.
3. Love/affection/and the sense of belonging (social interaction).

4. Self-esteem (self-confidence, prestige, level of importance).
5. Need for self-actualization.

The extension agent or facilitator must be aware of the fact that the farmer is a responsible person who is a parent, family man or a group leader and performs many social roles. He/she has to go to the market, attend funerals and take part in festivals and religious activities. The farmer is proud of his personality and expects to be respected for what he/she is as well as the roles played in the community. He/she is a very busy person who is time-conscious. These issues are important when considering the basic principles of adult teaching and learning.

*Adult Capacity for Learning*
Adults have great capacities to learn. They can learn issues which are very relevant to their needs. They can also learn effectively in a congenial, friendly or familiar atmosphere.

**Principles of Adult Learning and Teaching**

According to Knowles (1970), adult teaching refers to the "art and science of helping adults learn". It is aimed at opening opportunities, for their self-development, change in attitudes, updating of their skills and acquisition of new knowledge. According to him, adults are prepared to learn but in a friendly environment.

Many people, who are involved in teaching adults, have not received any form of training in the techniques of training adults. They specialize in their own fields of professionalism such as the basic sciences, agriculture or nutrition. They, therefore, lack adult education skills, which would enhance their performance as adult trainers.

Some adult educators rely on the background techniques of formal education that they have acquired to teach adults. Unfortunately, these techniques are woefully inadequate for adult education work. The extension agent (trainer or facilitator) needs to acquire some basic skills that will enhance his/her work in the field. This is similar to the condition of a tadpole and frog (*see* Fig. 3.1). A tadpole lives in water like a fish but when it matures and changes into a frog, it has to acquire new skills that will enable it to successfully live its new life on land.

*Fig. 3.1:* Like the tadpole and the frog, extension officers need to acquire a whole new set of skills in order to become good trainers and facilitators.
Source: Adapted from Hope and Timmel 1984.

Adult educators like Freire (1970), Bown and Tomori (1979) and Knowles (1970) describe and recommend some basic principles for adult teaching and learning as follows:

## Principle 1
### *Adult learning is a voluntary activity*
Adult learners are voluntary learners and have a variety of needs which they want to satisfy. If the needed motivation is not provided to achieve their aim, they withdraw from participation in the learning process. Opportunities and a variety of motivational techniques should, therefore, be used in a friendly atmosphere to keep participants regular and alert at training sessions.

## Principle 2
### *Learning is a personal experience*
Adult learning is primarily a personal involvement in and commitment to the learning process. Individuals must willingly devote their time and resources

to the process. It is only then that effective learning can take place. The learner controls the learning process and assimilates what is to be learned. Self-concept and dignity enable the adult to make decisions to develop. The ability to learn is, valuable and must be promoted by the facilitator.

The facilitator creates a favourable and friendly environment in which learners explore and discover the meaning of the world around them. They must be allowed to fully participate in the learning programme from the developmental stage through implementation to the evaluation stage. Individuals must take responsibility for learning and should not be spoon-fed.

Adult learners feel proud if they are involved in the learning process and, therefore, actively participate in it. They continue to participate in the training sessions if they believe that the programme is initiated by them and they have the opportunities to participate fully. Generally, no one person can learn for another, each person has to experience the learning process in order to obtain and appreciate the full benefit of learning. Learning is an experience which individuals have to go through consciously.

**Principle 3**
*The content of adult teaching must be of immediate relevance to beneficiaries*
Adult training is not merely a mechanism for producing literates. It produces people who would be able not only to learn to live happily but also to earn a living. Adults, therefore, readily participate in programmes that are of immediate relevance to them. The programme should include issues which learners should either be worried about, interested in, hopeful for, fearful or angry about. Such issues are closely linked with their emotions. They are, therefore, easily motivated when issues affecting their emotions are discussed since emotions enhance motivation. Adults learn best when the context of the training is close to their own jobs or tasks or real life situations. For example, the problem of pest damage to crops must be discussed with farmers immediately it occurs on the farm. Extension agents must always work closely with farmers on farming problems as they emerge. They should identify the priority problems and needs of farmers which they must all discuss and tackle together. Farmers become easily bored, and more especially so, if what they learn will not be useful to them immediately.

## Exercise 3.1 The importance of relevance in issues discussed with farmers

*Purpose:*
The purpose of this role-play is to demonstrate the importance of addressing issues that are of direct relevance or concern to farmers.

*Materials Required:*
Newsprint, coloured markers or crayons.

*Procedure:*
The scene is a group of six worried farmers gathered in their tomato farm to await the arrival of the village extension agent called Mr. Foli.

As the farmers await the arrival of the village extension agent, they discuss their fears and concerns about the continuous destruction of their tomatoes by pests.

The farmers are excited at the arrival of Mr. Foli in the hope that finally, a solution will be found to their problem. They appoint a spokesman to explain their tomato pest problem to Mr. Foli.

Surprisingly however, Mr. Foli does not address the pest problem but continues to preach to them about the kingdom of God. His attention is drawn to the pest problem but he becomes more aggressive in his preaching. The farmers show signs of frustration and begin to move away. At this point, the facilitator stops the role-play and discusses it using the following sequence and questions:

1. Discuss and list on the newsprint the problems that are of major concern to the farmers.

2. List on the newsprint issues discussed by Mr. Foli when the farmers complained to him about their problem.

3. Compare the list of farmers' problem with the subject of Mr. Foli's speech to them.

4. Why were the farmers frustrated by Mr Foli's speech?

5. How did the farmers react to Mr Foli?

6. What should Mr Foli have done to secure the attention and interest of the farmers?

7. Discuss the importance of the relevance of a subject in talking with farmers.

**Principle 4**
*Adult learning is an evolutionary process*
Adults have developed and nurtured different types of skills, knowledge, and experiences over time. It is, therefore, difficult to radically change their lifestyles and attitudes. This is the basis for the popular saying that *an old dog cannot learn new tricks*. Discarding old methods of doing things and adopting new skills are usually difficult, especially, for adults. Adult teaching, therefore, demands a lot of patience. Extension agents need the patience of missionaries to work with their learners. There should be constant visits and workshops for the learners to adopt the new technologies imparted. The process is always slow. When a friendly atmosphere is created, adults acquire skills slowly but surely. The facilitator should not rush adults through the learning process.

**Principle 5**
*Problem-solving techniques are suitable and necessary in adult teaching*
Adult learning is a process of problem identification and problem solving. It is, therefore, a problem-centred rather than subject-centred activity. No adult learner is completely ignorant of his/her work or life. Thus, adults want to identify problems and improve performance. Each adult has a unique package of skills and experiences in solving problems. Contributions from individuals can help solve even the most difficult problems. Facilitators must avoid using the "banking approach" whereby knowledge is supposed to be transmitted by the extension agent to the farmers. This is a teacher-centered technique. In this system, the teacher, *the supposedly living encyclopedia,* "fills" the learners with ideas that may not even be relevant to the needs of the learners. Adults may feel threatened to cope as ideas are forced on them. If ideas are forced on adults they are domesticated but not liberated.

With the problem-posing approach, farmers are recognized as intelligent people. Right from the beginning of the programme, they are taken as principal partners with the extension agent and researcher. What the facilitator does is to create the congenial atmosphere for learners to feel proud to participate in the process. Questions are asked and problems are posed for all to tackle and find answers and solutions. Learners rather than facilitators are to identify and solve problems. They are motivated and given opportunities to discuss issues freely. They ask questions, make comments and answer questions. This process then creates the right climate for farmers to raise questions that help them to identify their problems, the root causes of these problems and find solutions to them. The whole process is referred to as **experiential/participatory learning**. Farmers can more easily recollect what they say or do than what they hear or see demonstrated.

This problem-posing process exposes farmers to the reality of their farming conditions and of the world. It develops their critical consciousness to question conventional farming methods and adopt new farming techniques to produce more food. Farmers are thus actively involved in describing, analyzing, making decisions and taking actions to change their situations and conditions. Adult teaching is, therefore, a process in which learners and teachers work together to find solutions to problems.

## Principle 6
*Learning is a co-operative and collaborative process*
People prefer to learn or teach independently. But as the popular saying goes, *"two heads are better than one"*, (that is, provided both heads are good!); the facilitator should demystify the notion that he is an authority so that participants should take him/her into partnership in the process. The prime concern of the facilitator is to promote the participation of all in the learning process.

Adults learn best in an atmosphere of active involvement and participation. Building on farmer's experience, the facilitator must make them feel important and necessary. They also learn best when they have some control and ownership over what they learn. The facilitator also learns from the experiences of participants.

Learning in groups or involving many people in the learning process enhances the education process. Many adults prefer learning in groups because groups provide supportive environment for them to learn. Teaching and learning in a friendly group environment enhance keen competition among farmers to learn and apply the skills acquired. Farmers learn from one another and share the stock of experiences they have for the improvement of the group as a whole. Friendly relations are built amongst members of the group. Learning together is education.

Moreover, in view of the fact that adult education associations have limited resources for their operations, it is not viable to organize individually-based learning programmes. Many individuals can be reached through collaborative programmes. The involvement of agricultural agencies like Sasakawa Global 2000, Ministry of Food and Agriculture, Ghana Organic Agricultural Network, (GOAN), or World Vision International and GTZ field programmes or people interested in the promotion of agriculture, enhances a wider agricultural development. A network of all existing agricultural agencies must be built in the community so that all available resources can be pooled together to enrich training programmes. Networking will also reduce duplication of efforts and promote maximal use of resources. *Cross the river in a crowd and the crocodile won't eat you* is a Madagascar proverb which tells us that working together in a group is highly beneficial to the individual members of the group (*see* Fig. 3.2).

*Fig. 3.2:* Working together in a group is highly beneficial to the individual members of the group.
Source: Adapted from Hope and Timmel 1984.

**Principle 7**
*Learning leads to changes in the lives of those involved*
The main objective of every educational programme is to improve the skills, knowledge, attitudes and practices of people so that they can actively participate in the development process. Extension agents, trainers and facilitators demonstrate what they teach and become role models for their beneficiaries. Therefore, changes are expected to occur in the lives of learners.

More importantly, beneficiaries must show evidence of change in their relationships with others and discard some of the outmoded cultural ways of life and work. Farmers should learn to plant and maintain improved varieties of crops, know when, where and how to obtain credit facilities and relate more cordially with their neighbours. They must be integrated into modern systems of production. In effect, there should be marked improvements in the lives and work of both the extension agent and the farmers.

An important issue that extension agents should note is that teaching adults is quite different from teaching children or young people. In the former, the experiences of the adults form the basis for education. The attitudes of adults are already formed and they have a large stock of experiences. The extension agent only provides the framework for farmers to think and improve upon their capabilities. The extension agent raises questions and through proper direction, farmers provide answers to improve their performance.

**People's Values**

Facilitators should not however, assume that these principles will work automatically and perfectly well in all situations. A particular problem which faces many leaders and extension agents is the variety of human characters that they have to work with. Some leaders assume that all the people they work with are generally the same. But a leader must be aware that because of differences in upbringing in various environments, individual aspirations and behaviour are different. People have different values, which influence their behaviour.

The story of Serwa, adapted from Stapley (1982), can be used to demonstrate how different values and situations influence individual behaviour.

## The Story of Serwa

Once there was a girl called Serwa. She was 19 years old and was very beautiful. She was also very poor. She lived in a village on the bank of a big river. Serwa was engaged to be married to a young man called Mensah who lived in another village on the opposite side of the river. The river was wide and fast flowing and contained many wild crocodiles.

One day, Serwa heard that Mensah was very ill and might even die. She became very anxious about Mensah. She loved him very much and wanted to go and be with him because he was sick and might die.

So she went down to the river where there was a ferryboat. A man called Yaro rowed the ferryboat. When Serwa said she wanted to cross the river, Yaro asked her to pay ¢500. Serwa said that she did not have ¢500 at that time but she could pay him later. Yaro refused. Then Serwa pleaded with him to take her because Mensah her fiancée was very ill and might die and she would lose him. Yaro again refused. Then Yaro later said that he would take her across the river only on one condition; that Serwa should sleep with him first.

Serwa was very upset about this and went back to her village wondering what to do. On the way home she met her cousin Baba to whom she narrated the story. "That's nothing to do with me" he replied her. "That is your own problem. Do not involve me in it. I do not want to have anything to do with it". Then Baba went off leaving Serwa confused.

Serwa did not know what to do. She hated the idea of sleeping with Yaro, but she loved Mensah so much and thought that she might never see him again if he was to die of his illness. She had to get across the river somehow to see Mensah. So finally, she went back to Yaro and slept with him. Then he took her across the river and she rushed to Mensah's house.

At Mensah's house Serwa nursed him and looked after him. Soon Mensah felt better and was out of danger of dying. After some time, Mensah asked Serwa how she crossed the river and where she got the money to pay the ferryman. Then Serwa told Mensah what had happened between her and Yaro. Mensah was furious. He shouted at Serwa and insulted her calling her rude names for having slept with Yaro. He told her that he would never marry her and that she should get out of his house forever.

Serwa went sadly away down to the ferry again. On the way she met a neighbour called Nana. She narrated everything that had happened to Nana who was very angered at what he heard. He immediately rushed to Mensah's house, dragged him off his sick bed and beat him up very badly.

## Exercise 3.2 Individuals think and behave differently according to their perceptions of a situation

*Purpose:*
To demonstrate that perceptions and attitudes of people influence their behaviour patterns.

*Materials required:*
Copies of the Story of Serwa, Newsprint.

*Procedure:*
1. Make copies of this story of Serwa and distribute to training participants to read aloud in class.

2. Each participant should then rank the five characters, namely Serwa, Yaro, Baba, Mensah and Nana, according to the seriousness of the crime each of them committed.

3. Individual rankings are then compared with those of others.

4. Draw lessons from the rankings to show that people think, speak or behave differently.

5. Write the major lessons learned from this story and the rankings on newsprint, discuss them and paste the newsprint on the wall of the classroom.

The story of Serwa can also be used to help participants learn how to solve human relations problems.

1. Participants are divided into groups of four or five to agree on a common ranking. Each participant should try to persuade others to agree on his/her ranking until the group agrees upon a common ranking.

2. Later, the larger group can be divided into five smaller groups each adopting a name of one of the characters in the Serwa story. Each character group then discusses what could have been done to improve relations in the story.

3. Suggest how the characters in the story would have behaved to improve relations between them. Let each group state the value each of the actors had and how individual values influenced their actions. Lessons are drawn on how individuals can improve their lives and relationships.

CHAPTER 4

# Approaches and Methods of Non-formal Education

The objectives of this chapter are:

1. to illustrate the concept and importance of non-formal education;
2. to explain the methodologies that can be used in adult training sessions; their characteristics and the most appropriate situations to use various methods.

## Forms of Education

There are three major forms of education: Formal, Informal and Non-Formal.

The formal system is the classroom type of education which is conducted by teachers. It is graduated from one stage to another. A common syllabus and training curriculum are drawn up for all the schools in the system. Academic certificates are awarded at the end of the programme after students pass prescribed examinations. These certificates are very important for self-confidence and in the search for employment.

The informal system refers to all learning that takes place by chance. It can be through conversation, observing what others do, viewing television or general reading. It is not planned.

The non-formal system includes all organized education outside the formal system. It is organized for specific target groups like farmers, artisans, extension workers and women groups. Its programmes are tailor-made to suit the target groups and they are organized on short-term basis. Certificates of proficiency or of attendance are awarded at the end of such programmes. However, such certificates are usually not submitted as academic qualifications for seeking employment, although they can confer advantages in the selection of candidates for a position.

## Advantages of Non-formal Education

The world is changing very fast, and the present electronic age presents new demands that are varied and complex. Adults cannot rely only on the

## Approaches and Methods of Non-Formal Education    35

skills they have acquired through formal and informal education to meet the demands of modern living.

Many adults are unable to spend long periods in school to acquire new skills. Non-formal education programmes, therefore, become particularly useful to help people meet their social, economic and political obligations. Socially, non-formal education can prepare citizens to know their civic rights and responsibilities through programmes organized by agencies like the National Commission for Civic Education. People who attend such educational programmes are assisted to understand issues related to the development of their communities and so they can participate in community-improvement activities. Furthermore, non-formal education provides opportunities for people to share ideas and experiences. They learn to relate well with others and improve their communication and human relation skills. They make new friends and meet older and more experienced people at learning centres. The latest rumour and jokes are shared, and problems are discussed and solved.

Economically, non-formal education provides manpower at all levels of the socio-economic ladder. Through non-formal education, illiterate farmers learn to read, write and create their own news. They are able to read names and directions of farming inputs and equipment. They learn new methods of production, including the planting of high-yielding varieties. They also learn the prevention of post-harvest losses, building of storage facilities, adoption of marketing strategies, where and how to obtain credit facilities. Farmers share work experiences, discuss the latest inputs in the system, adopt new strategies to earn their living. The youth are also trained to learn a trade or to become farmers.

Non-Formal education provides farmers with the opportunity to organize and join co-operatives so that they can have a common front to fight for their rights and for good prices for their products.

Politically, non-formal education prepares individuals to take active part in the administration of their communities. Farmers learn voting procedures, the right and responsibilities of individuals in the society, the functioning and advantages of joining clubs and parties such as village IPPM Clubs. They learn issues relating to the political system in their country, including power structure, the new constitution and the District Assembly systems. Laws of state and by-laws of District Assemblies are explained to participants so that they conform to the national system of laws and by-laws for the maintenance of peace. The general regulations governing farmers' organizations are explained to farmers in non-formal education sessions.

A variety of approaches in non-formal education illustrated by Bown and Tomori (1979) and WARDA (1991) are now briefly described.

## Non-formal Education Techniques and Methods

The facilitator of a non-formal education programme is responsible for helping adults learn. He/she must understand and use various methods to enhance the learning process and sustain adult interest in the programme.

The terms technique and method are sometimes used interchangeably. Techniques are loosely used here to refer to how the adult educator handles teaching and learning aids and how he/she relates to or handles learners so that he/she will teach effectively. Methods are used loosely to refer to the packaging of materials and the procedures adopted in achieving educational objectives. Some of the essential features of techniques and methods are summarized as follows:

### *Techniques*
The techniques the facilitator will need to use are:

1. Prepare notes and materials well in advance before you meet learners.
2. Train learners to enter the classroom or centre in an orderly manner.
3. Discuss previous lessons with learners.
4. Set objectives for lessons on a daily basis.
5. Introduce some humour into class discussions.
6. Go through the lessons gradually and do not rush.
7. Position yourself where each learner will see you.
8. Do not turn your back to learners when writing on the board.
9. Talk to learners but not to the board when you are writing or pointing at the board.
10. Use pointers with the right hand to illustrate issues on the board or elsewhere.
11. Collect learning materials in an orderly manner.
12. Moderate your tone according to issues discussed.
13. Watch the mood of learners when teaching and avoid teaching when they are tired.
14. Use a variety of teaching methods.
15. Train learners to leave the centre in an orderly manner.

## Methods

### Lecture

The lecture method is the most popular traditional method in the transfer of knowledge. In non-formal education, it should be used sparingly. This is the method used to teach some technical issues or concepts. It is very useful for large numbers of learners when the number of facilitators is limited. All learners in the same class are taught the same topic together.

### Discussion

This is the method recommended for non-formal education work. Participants must be encouraged through the ingenuity of the facilitator, to discuss topics. The initiative of learners is built. Learners feel comfortable to raise issues. They then accept decisions as their own. This method has many advantages since farmers themselves are involved in the learning-teaching process. Confidence is built among learners as they discuss common issues together. Collective action is taken to implement developmental activities.

### Socratic Techniques/Question and Answer Technique

This method, popularized by Socrates the Greek philosopher, is related to the discussion method. Here, the facilitator asks questions mainly to direct the course of discussion. The, *Who, When, Which, Why, What, Where* and *How* questions are asked till the root cause of the problem is identified and thoroughly discussed and solutions found for addressing the issue. This method is also commonly known as the 6 Ws and 1 H method or simply as the "seven friends method".

### Case Study

The case study is the presentation and analysis of an event or scenario involving a specific problem that has happened or could happen. A case study should be realistic and should illustrate a situation occurring among people. The purpose of the case study is to make the case as practical as possible so as to help people to acquire skills and knowledge to solve problems. Participants must know how the difficulties develop and find ways of tackling them. Case analysis should, therefore, be comprehensive and flexible to allow effective participation. Opportunities are given to participants to reflect and explore alternative situations to solve problems.

*Demonstration*
This involves practical work when the facilitator physically participates and shows how things are done. Others take their turn to practise what has been demonstrated. This ensures that all those who participate get the message through "learning by doing". It is one of the best ways of teaching manipulative skills.

*Seminars*
In this case, outsiders who are experts in their field of specialization are invited to present papers. It makes the programme very rich since presenters raise a variety of issues. Some of the questions, for which the extension agent seeks answers, are fully discussed under the guidance of the experts.

*Excursion/Field Trip*
Members of a group can visit farms or agricultural stations or institutions to know about what others in the same field are doing. New methods and ideas of cultivating crops may be acquired during an excursion.

*Colloquy*
Colloquy is a method used to involve both experts, who may be two or more, and learners in an open discussion of issues related to their work. Both partners ask and answer questions and exchange ideas to find the most profitable ways of doing their work.

*Role-Plays/Drama*
Role-plays are simplified and very brief forms of drama directed at analyzing and finding solutions to problems. Situations and specific problems are acted by participants. They see their problems mirrored to them and they are excited to discuss and find solutions to them. Role-plays are very effective and exciting ways of involving people in the learning process.

*Stories, Proverbs and Songs*
Learning methods in traditional African societies include Ananse (spider) stories, proverbs and songs. In these techniques, values are exposed to learners through these media. Stories and songs can be used to facilitate learning as they are valued in society. It is, therefore, important that the facilitator selects some pertinent stories, proverbs and songs to enhance the learning process.

## Games

Local games related to particular lessons can be introduced to stimulate learners to learn and find solution to problems. Games are exciting and motivate participants to continue to learn. They expose learners to the realities on the ground (e.g. Win-Lose Game). The facilitator is not engaged in any sort of formal teaching or supervision.

## Counselling

This is done on individual basis to help learners develop regular learning skills. It enables individuals to spend enough time with the facilitator to solve some personal problems.

## Individual Teaching/Tutorial

Like counselling, tutorial is done on individual or small group basis to give the chance to the learner to seek clearance on issues he/she could not understand in the class. New directions for studies are given to the learner. Each learner is provided with learning aids to assess his/her own progress.

## Small Group Work

Learning becomes very effective when people work in small groups. Every member of the small group has ample opportunity to participate in the programme either orally or practically.

There are two ways of grouping learners

1. *Ability Grouping:* This is the method whereby participants of the same ability or interest are brought together to solve a problem. Participants are ranked and grouped into ABC or more.

2. *Mixed Ability Grouping:* This grouping is done by any simple or random method to break the large class into groups for easy management. The ability or interest of learners is not used as a criterion for the grouping.

## Activity Method

In the activity method, learners undertake some activities such as drawing, dancing, singing, reading or experiment. The idea is to give participants the opportunity to discover things for themselves through these activities. It enables participants to apply theories learnt to solve problems. Practice and practical lessons are components of this method.

*Project*

In the project method, a theme about the subject is chosen for in-depth studies. Participants may be grouped to study an aspect of the theme or individuals work on different aspects or all aspects of a subject.

*The Dalton Plan*

This method is the assignment method, which is named after Dalton, a town in America, where the method originated. Students are left free to use any method to arrive at answers to respond to questions asked. A variety of methods can be used such as reading, group discussion and research. But individuals come out with their own reports.

*Discovery Learning*

Discovery learning means a situation whereby the facilitator exposes learners to the problem or solution while learners are requested to find the solution or the problem respectively. The main issue here is that learners work on their own with little guidance from the facilitator to discover answers. In any of these cases, the facilitator provides the necessary illustration, asks leading questions and provides the learners opportunities to ask questions.

*Other Methods*

Other methods that can be used to bring messages home to farmers are debates, films and Interactive Video.

In all these techniques and methods, the facilitator needs to perform some maintenance functions to achieve the teaching/learning objective. These include listening, gatekeeping, encouraging, relieving tension and harmonizing. By gatekeeping, the facilitator invites everybody, especially the quiet ones, to participate in the discussion. Dominance of few people or one person in the discussion is avoided. Time and number limit on participation by individuals may be set. Any other procedures to keep open multi-communication avenues are introduced by the facilitator and the group (Pretty *et al.*, 1995)

By encouraging, we mean that the facilitator creates a friendly environment. He/she is warm, affirming the responses of participants and showing appreciation of their views even when disagreeing.

By mediating or harmonizing, the facilitator clears up confusions, reconciles differences in opinion and helps those in conflict to understand others' views.

# Approaches and Methods of Non-Formal Education 41

A well-timed joke is cracked to kill boredom or relieve tension. The facilitator can ask for a "cool off" period by introducing energizers or a short break.

During all sessions, the facilitator should let the group remain on target, giving concrete illustrations and directing participants towards taking appropriate actions to solve problems.

**Qualities of a Good Facilitator**

For the appropriate use of these techniques and methods, the facilitator needs to have some special qualities, including the following:

1. knows the awareness levels and educational backgrounds of his/her learners;
2. controls his/her emotions;
3. motivates and sustains the interest of learners in learning;
4. knows and uses a variety of methods and techniques in teaching;
5. treats all learners equally;
6. is committed to his/her work;
7. is able to evaluate learners accurately;
8. adopts appropriate communication skills;
9. promotes full participation of learners in class activities;
10. knows how to prepare, and use teaching/learning aids;
11. continues to learn;
12. has confidence in himself/herself;
13. creates a friendly learning environment.
14. is honest and committed; and
15. is punctual and regular.

CHAPTER 5

# The Communication Process

**The objectives of this chapter are:**

1. to explain the communication process;
2. to demonstrate the importance of effective communication in the facilitation process;
3. to explain some barriers to effective communication;
4. to highlight some of the ways to improve communication among farmers.

## The Importance of Communication in the Facilitation Process

Communication is very important for inter-personal relationships and plays a vital role in farming situations. Where farmers work in groups and researchers and extension agents try to introduce new farming innovations, communication becomes the main vehicle for dissseminating information. However, it must be stressed that information by itself is not enough to change the behaviour of farmers. Other factors come into play because many things tend to interfere with the accurate transfer of information from one person to another.

Why is effective communication so important in the organization of Farmer Field Schools? One major reason is that good communication is the key to effective group dynamics. Members of any group must communicate with themselves and understand one another in order to best make collective decisions and to effectively organize themselves for community development.

We have already seen that a friendly and favourable learning climate is characterized by every member of the group having absolute freedom to speak and be listened to and to engage in meaningful dialogue with others. If communication between members of a group or between trainer or facilitator and farmers is poor or ineffective, people tend to misunderstand themselves, develop wrong or unintended perceptions which block the learning process. Sometimes, this situation may lead to conflict. Good communication fosters good working relationships among workers and eliminates suspicion, unrest and friction. It is a fundamental and vital process in the existence and survival of societies and organizations.

In agriculture, there can be no diffusion of information or innovation and technology transfer cannot occur without some kind of effective communication.

## Interpersonal Communication

A simple definition of communication is that communication is the process of sending and receiving messages or knowledge. In literature, or in general conversation, we come across various definitions of the term communication because different groups define communication in relation to their work and environment. We believe that adopting the simple definition of communication given above is appropriate in our work with farmers. Communication is a process because it is a continuous and on-going activity.

According to Finlay (1994), the communication process is a set of actions or changes that bring about a specific result. It involves interaction between a sender, the message and a receiver. So the elements of communication can be explained as follows:

1. The sender — a person who chooses words and puts them together (encode) in a way he/she thinks will carry some meaning to another person. The sender is described as the encoder of the message.

2. The message — that is a set of ideas put in the form of symbols and transmitted through a medium such as the human voice, by visual image, by working through some channel; person to person, telephone, radio, television, electronic media such as e-mail and the internet, and traditional media such as storytelling, plays, songs, dance and proverbs (extended channels).

3. Receiver — another person, who picks up the message as sound or light impulses, converts them into words and assigns some meaning to the combination of words (decode). The receiver is said to be the decoder of the message.

The process and requirements for effective interpersonal communication can be presented graphically as in Fig. 5.

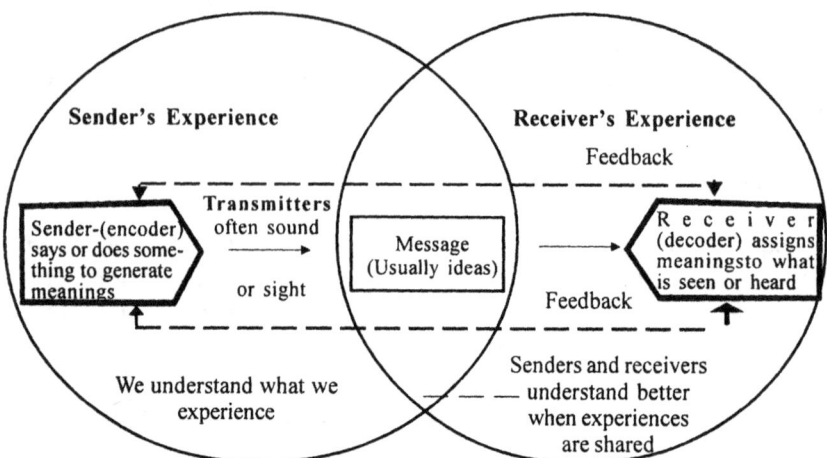

***Fig. 5:*** *Requirements for the communication process: (adapted from Lionberger and Gwin (1992) Communication Strategies: A Guide for Agricultural Change Agents.*

## Directional Ways of Communication

Three main directional ways of communication are generally recognized. These are:

1. One-way communication — in which one person (sender) sends a message to another person (receiver). This can be illustrated like this:

sender ⟶ message ⟶ receiver

2. Two-way communication — in which one person is talking and others are answering; the sender thus makes provision for receiving feedback of the message. This is illustrated as:

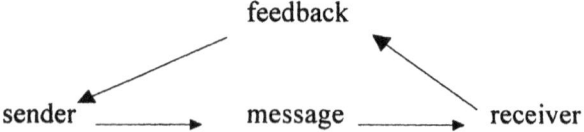

In the two-way communication, feedback can be given in both verbal and non-verbal forms. Some responses/feedback are realized immediately while other messages are not immediately recognized. Response/feedback may also take the form of a change of behaviour or action.

3. Circular communication/multi-communication — in which the sender and receiver change places, asking and answering questions back and forth until both of them feel that they understand each other. In large groups, all members participate in the process. This may be illustrated as:

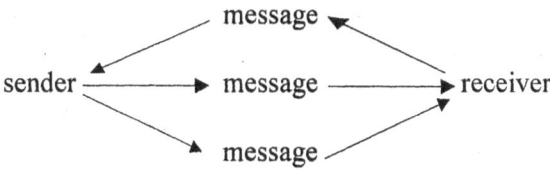

Communication is generally a two-way affair as both the sender and receiver are engaged in playing dual roles; both receiving and sending feedback throughout the communication process. It cannot always be assumed that during the process of communication, the receiver attaches exactly the same meaning to the message as the sender intended. This is because the sender attributes certain meanings, based on his/her experience and environment, to the words used to send the message, while the receiver also attributes his/her own, and perhaps different meanings, to the words received, based on his/her own experience and environment.

Clearly then, in communicating, especially with farmers, it is important to use very simple and clear words and expressions which fall, as much as possible, within the common shared experiences of both sender and receiver.

**Forms of Communication**

We can distinguish between different forms of communication as follows:

1. *Monologue* — occurs when one person speaks and others just listen without responding. Examples of monologue forms of communication

are classroom lectures, church sermons, presidential addresses and political campaign addresses.

2. *Dialogue* — occurs when people exchange ideas, for example, during focus group discussion by people with similar interests, background or a common objective.

3. *Concurrent monologue* — occurs when two or more people are speaking and saying different things to each other — this is a very defective form of communication because neither of the parties is listening!

4. *Narrator sickness* — occurs when one person talks too much and continuously to the annoyance of others who are not given a chance to speak and contribute to the communication effort. This is another very defective form of communication.

### *Exercise 5.1 Monologue and dialogue forms of communication*

*Purpose:*
This role-play is designed to demonstrate patterns of communication and an analysis of the main features of directional ways of communication.

*Materials Required:*
Six participants, large sheets of newsprint, colored markers, cellotape, scissors.

*Procedure:*
1. The six participants should pair up to make three pairs: A, B, and C.

2. The pairs should be of the same sex and age range so as to eliminate any form of bias or prejudice.

3. Each pair should perform a role-play, described below, while other participants watch and take note of how effectively the people are communicating.

4. Conduct briefing of the pairs as follows.

## The Communication Process 47

**Pair A.** Two people meet. They do not exchange greetings. One starts talking and gets so excited that he/she does not pay any attention to the other person. The other person tries very hard indeed to speak, to ask a question or even to stop the other person speaking but is unable to do so. He/she becomes frustrated and gives up and remains completely silent. Stop them when this point is reached.

**Pair B.** The two people meet and exchange greetings briefly. Each one starts speaking, and saying completely different things to the other, no common topic being discussed. Neither is listening to the other and both are talking loudly at the same time. Stop them when you are sure that participants have realized the point that the role-play is illustrating.

**Pair C.** The two people meet and spend some time exchanging greetings warmly. They begin to ask questions about each other's interests, especially about the performance of their farms, about the last harvest and the pattern of visits by the local field extension agent. They listen patiently to each other and ask several questions, share ideas, talk about current news and views. Finally, they thank each other, shake hands and bid themselves farewell.

5. Give actors five minutes to act this role-play.

6. At the end of the role-plays, the six participants should go back to their places in the group for further group work and general discussion.

7. Divide the group into three working teams to assess the level of communication in each role-play. Supply sheets of newsprints and coloured markers to each working team to write down their observations and answers to the following discussion questions.

   i. What happened in play A?
   ii. What happened in play B?
   iii. What happened in play C?
   iv. How would you describe the pattern of communication in plays A, B, and C?
   v. How often do these patterns happen in real life?

vi. In which play did you note effective communication? Why?
vii. What can we do to ensure that there is effective communication between people?
viii. List about four of the things we can do to improve communication.

8. The whole group should now come together for general discussion.

Each working group should present the answers to the questions followed by a general discussion of the observations and answers to the questions. At the end of all three presentations, the facilitator should summarize the main points agreed upon and write these on a newsprint to be pasted on the classroom wall. These would then represent the group's Guidelines to Good Communication.

**Channels of Communication**

The channels of communication between sender and receiver are the five senses, namely Hearing, Seeing, Touching, Smelling and Tasting.

**Types of Communication**

Two major types of communication are important for trainers and facilitators. These are oral communication and non-verbal communication (Kunczik 1984).

*Oral communication*
Oral or verbal communication includes the spoken word or speech and written messages. Written communication refers to reports, press release, letters, memos, notices, questionnaire and journals. In this case, speech is recorded in written form. Oral communication refers to situations where speech is involved, for example as in conversations, lectures, and instructions that are delivered by speech.

Research has shown that people spend over 80 per cent of their communication time speaking aloud. This is true for most trainers and facilitators because they are always in personal contact with community members to whom they speak most of the time. To communicate effectively, it is important for trainers and facilitators to develop good verbal communication skills. Verbal communication is a very effective way to communicate because it allows for immediate feedback. It is more personal and encourages a feeling of personal involvement. It places emphasis on the human element in community work and the physical presence and voice of the Trainer or Facilitator.

*Non-verbal communication*
Non-verbal communication occurs in many forms including:

1. Appropriate posture,
2. Meaningful and realistic gestures,
3. Friendly eye contact,
4. Miming,
5. Demonstration,
6. Body language,
7. Open facial expression,
8. Adequate proximity that shows respect for personal space,
9. Cheerful mood,
10. Modest dressing,
11. Respect for others,
12. Statistical tables,
13. Charts,
14. Diagrams,
15. Photographs,
16. Fire/drum,
17. Whistling.

**Routes of Communication**

There are three main routes of communication. These are the vertical, lateral or horizontal and diagonal routes.

*Vertical.* Vertical route of communication is the top-down communication where instructions come from the top to workers who implement the instructions.

*Lateral/Horizontal*
In lateral or horizontal cases, communication is among people who operate at the same level. This is concerned with the flow of information and not

with the flow of instructions from a higher to a lower authority.

*Diagonal*
Diagonal communication involves communication with senior colleagues.

**Barriers to Effective Communication**

Communication is a complex process. Sharing information or ideas for everyone to understand the exact meaning is not easy. Many factors prevent people from understanding messages in exactly the same context as they are sent. To describe this phenomenon, it can be said that communication is affected by "noise" or "disturbance." Disturbance may be a learner factor, a trainer/facilitator factor an environmental factor, or a combination of these factors.

Some examples of noise or disturbance factors are:

1. Financial difficulties,
2. Domestic problems,
3. Health problems,
4. Hunger and poverty,
5. Bias,
6. Outlook and cultural background,
7. Gender perspective,
8. Illiteracy,
9. Social problems such as death of relatives or friends.

The background of the receiver is a major factor that influences understanding of a message that is communicated. It is not safe for trainers/facilitators to assume that everyone in a group is receiving messages in the same way. Each person has personal characteristics, for example, sex, educational background, marital status, age, religion, culture, ethnic origin, occupation and political affiliation which influence the interpretation of the message being communicated. People receive and understand messages in different ways. This is nobody's fault; the ability to receive and understand messages results from marked differences in backgrounds and experiences, over which we have very little or no control whatsoever. Education may, however, attempt to bridge the gap in understanding issues communicated.

Trainers/facilitators should, therefore, endeavour to identify differences in people based on their backgrounds and life experiences, their attitudes, feelings and behaviour. By understanding the persons receiving the message, the trainers/facilitators can arrive at a common denominator that will help them to cordially relate and communicate directly and more effectively with listeners. Disrespect to persons or to their culture or any other personal characteristics will be resisted and this creates problems in the communication process.

Perception is the way a person understands issues and symbols. The socio-economic background (sex, age, education, culture/religion, environment and marital status) influences the perception of the sender and the receiver in the communication process.

*Noise* is any variable that interferes with the communication process. It can be **physical noise** such as people shouting. Human problems such as poor eyesight, and hunger also prevent effective communication. **Physiological** noise can involve biological problems such as illness and weakness due to old age. Low levels of intelligence and individual levels of absorption can have negative effects on communication. **Psychological** noise can include fear of failure. **Social** noise may arise from poverty, ethnic conflicts, illiteracy and environmental degradation. If these conditions exist, communication is usually unsuccessful. Trainers/facilitators should consider these noise factors, confront the problems directly and seek ways of reducing them in order to communicate effectively.

*Information overload* is another situation that can constitute a barrier to effective communication. Information overload occurs when too much information is sent at any one time to a receiver. When too many instructions are received, difficulties in interpretation occur and this negatively affects understanding. The situation becomes more complicated when many items are mentioned or introduced at the same time. Human senses can only accommodate and interpret a certain amount of information at any given time.

The language used for communicating a message should be simple and clear in order to facilitate understanding of the message. Trainers/ facilitators should, therefore, choose their words carefully so that the words used for communicating messages are relevant to particular communities. The use of the personal pronoun "I" may create noise. Too many statements that begin with this pronoun may mean that the Trainer/Facilitator is mainly interested in himself/herself and ignores others. Words that make reference to the group such as mentioning the name of the community or individual names serve as a constant reminder that messages are directed at the group that should take action. When the group receives positive recognition, the communication process is enhanced. References to words or situations such as wars, battles, massacre of people, genocide, natural disasters, sex or issues which communities do not wish to remember can create barriers in communication.

Effective communication can be hindered by making references to difficult or sad episodes. Such references should, therefore, be avoided as much as possible.

Noise or barriers to communication can also occur in the sender, the message, the channel and /or the receiver. Individuals can produce noise or barriers to the communication process. The following may cause barriers.

1. Dress mode and appearance;
2. Unclear or unpleasant mannerism or speech;
3. Lack of eye contact;
4. Unnecessary and distracting body movements;
5. Discussing unpleasant or unrelated subjects;
6. Emphasizing unnecessary or lengthy details.

Noise can be internally generated (sender/receiver) or externally induced (the environment).

## Enhancing the Communication Process

Remembering the following points and trying to adopt them can improve the communication process.

1. Developing good listening skills to hear and understand others.
2. Using appropriate and simple language;
3. Speaking with a clear voice, but not shouting; adopting appropriate tone and voice pitch;
4. Displaying open and honest expressions;
5. Placing emphasis mainly on important and relevant issues;
6. Making speeches at an appropriate pace, not being too slow to bore people or too fast for listeners not to follow what is being said;
7. Choosing the appropriate tool for sending messages according

to the mood and background of the audience;
8. Appreciating people's dignity, personality and achievement;
9. Reducing all types of noises;
10. Keeping the messages as simple as possible.

Trainers and facilitators can develop techniques to overcome barriers to effective communication. They need to consider the ways they present themselves and modify their actions as appropriate. Trainers and facilitators must be aware of the effect of their words and of the impressions they provide to receivers. Words themselves may not be offensive but the way in which they are delivered and the mood in which they are received may make them offensive and, therefore, they can constitute barriers to communication. An important rule for effective communication is to keep the message short and simple. This is the KISS (Keep It Short and Simple) principle.

Listening is an important tool in the communication process. Listening goes beyond hearing. It involves the use of almost all the senses to interprete what is heard. It involves critical examination of the message and an appropriate response in the process. Poor listening skills lead to a breakdown in communication.

Listening is a skill which trainers and facilitators should acquire. Paying attention, concentrating with both mind and senses, being patient and actively listening to other people's points of view will convey real feelings and achieve successful communication. Trainers and facilitators should strive to be alert and sensitive to other people's situations in life. Communicators should have an open mind, listening ears, a sympathetic heart and a tamed mouth. Using communication tools effectively facilitates the communication process.

CHAPTER 6

# Interpersonal Relationships: Johari's Window

**The objectives of this chapter are:**

1. to explore the sense of self-awareness;
2. to share personal feelings.

**Introduction**

Johari's Window is a theory propounded by two American psychologists, namely Joseph Luft (Joe) and Harrison Ingham (Harry). The theory, which derives its name from the first names of the two psychologists, Joe and Harry (Johari) stipulates that each person is a mystery, partly known and partly unknown to himself and others. It illustrates that every person has four major aspects to his/her life. Each aspect is covered by a window namely the **free/open, blind, hidden and dark**. Each window represents a different aspect of human personality. The more the windows are opened, the more developed the person becomes.

This concept can be used to explain how individuals and groups can grow in self-knowledge and how they can improve their working relationships. Johari's Window (Fig. 6.1) also lays emphasis on listening and sharing of ideas and experiences to enhance individual and group development. We shall describe these windows using examples from role plays to illustrate the features of each window.

**The Free/Open Window**

The Free/Open Window is that aspect of life which is known to oneself and to others. It is a free window through which people know themselves. This window can be opened wider through discussion and conversation. For example, the open window can show the cheerful nature of an extension agent. Others can discuss this nature with the extension agent and encourage him/her to continue with the good work. However, the extension agent should not feel swollen-headed. Rather, he/she should continue to work

*Interpersonal Relationship: Joharis Window* 55

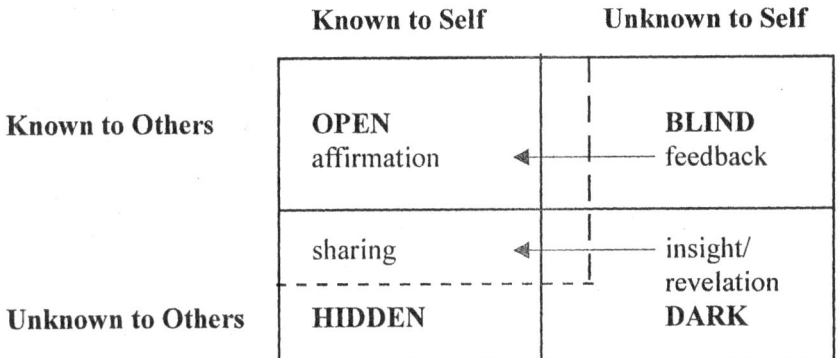

**Fig. 6.1:** *Johari's Window*
Source: Remodelled from Luft (1970). *An Introduction to Group Dynamics,* Mayfield Publishing Company.

cheerfully to keep this window open, and continue to win the support of farmers.

### Exercise 6.1 The Free/Open Window

*Purpose:*
The aim of this role-play is to demonstrate the presence of the open window and how it can be kept open.

*Materials Required:*
Newsprint, coloured markers or crayons, six participants.

*Procedure:*
One of the participants plays the role of the village extension agent, Mr. Agbotse, while the other five participants act as farmers.
   The five farmers sit under a tree awaiting the arrival of Mr. Agbotse who had promised that he would visit them that day. While waiting, the farmers discuss the cheerful and friendly attitude of Mr. Agbotse towards them and decide that they will encourage him to continue in this way because this attitude is helpful to farmers. Mr. Agbotse then arrives looking cheerful and immediately exchanges greetings with the farmers. He sits down with them and they start discussing the current food situation in the

village and country as a whole and later discuss some crop production problems. The atmosphere is very friendly and warm and everybody is pleased. The farmers then tell Mr. Agbotse that all the farmers in the community are very happy with him because of the pleasant manner he interacts with them. They encourage him to continue with this style of work and congratulate him for contributing very much to their success in farming. Mr. Agbotse responds that he is aware that he is a cheerful person and that people often make this comment to him.

Stop the role-play and discuss the advantages of the village extension agent being aware of his cheerful nature and how this attitude, the open window, enhances his work with farmers.

*Discussion Questions:*
1. Did Mr. Agbotse know that he had a cheerful character?
2. Explain your answer.
3. Did many farmers know that Mr. Agbotse was a cheerful person?
4. How can the open window be kept open or widened to enhance his performance?

**The Blind Window**

The Blind Window refers to an aspect of life which one is blind to. One does not know it but others know it. This aspect of the window can refer to the way an extension agent harshly relates to farmers, shouting when speaking to them. As a result, farmers start avoiding him/her by giving excuses. The extension agent does not know that shouting at farmers is bad and that, that method is not the best way to interact with farmers.

The best way to know about one's blind window is to ask for feedback on one's life and work. Feedback is the process of receiving and evaluating information about ourselves. It presents information on how our behaviour affects others. If it is well processed it becomes a yardstick to learn about ourselves and correct our intolerable behaviour. Feedback can be provided through verbal and non-verbal forms of communication.

Giving feedback is sometimes difficult because it may be critical to one's behaviour. The timing, the mood and the process of giving feedback must, therefore, be appropriate to the receiver. Feedback directed towards behaviour should be descriptive rather than judgmental. Although some types of feedback may be unpleasant, they can help field workers to improve their working relationships and their general performance. Some trainers find it difficult to be critical about their own behaviour and are not prepared to receive feedback. Learners also find it difficult to recognize their mistakes. They, therefore, block their own progress by avoiding issues that challenge their experience (Pretty *et al.* 1995).

Feedback is important for two main reasons: Firstly, constructive feedback reinforces the continuous good work of people, secondly it immediately addresses some mistakes which might have been repeated regularly. Self-reflection and group reflection are also good components of feedback enhancing identification of problems and potentials for improvement. Extension agents, trainers and facilitators, therefore, need to frequently ask for feedback from farmers and colleagues on their performance and working relationships. They must also give feedback in a friendly atmosphere.

### *Exercise 6.2 The Blind Window*

*Purpose:*
This role-play demonstrates that people are not aware of some aspects of their behaviour patterns. The play also shows that the blind window can be opened through feedback.

*Materials Required:*
Newsprint, coloured markers or crayons.

*Procedure:*
One participant acts as Fati, a village extension agent, who visits one of the villages in her area. Ten participants act as farmers who assemble to meet her. Fati is a very serious extension agent. She does not smile at farmers because she thinks that this will compromise her dignity and qualification and that being nice has no place in the work environment. When interacting with farmers, Fati shouts at them and looses her temper when farmers ask questions.

After meeting with farmers in this manner for about five minutes, the farmers stop asking questions, and begin to leave the meeting. Fati now realizes that the farmers are leaving but does not know the reason why. She then calls one of the farmers, Mr. Kwame, and asks him why the farmers were walking out of the meeting.

Mr. Kwame reluctantly tells her that many farmers complain about the way she conducts village visits and meetings. Farmers are distressed that Fati always shouts at them and does not permit free discussion with her but always looses her patience and insults them. Fati is surprised to hear this and responds that she is working very hard indeed, and doing her best to help the farmers but they do not show any appreciation for her work. Mr. Kwame continues by informing her that because of her behaviour, all the farmers in the community plan to avoid her and will no longer attend meetings or allow her to visit their farms.

Fati pauses for a moment and says, "so this is why farmers walk away from my meetings? Now I know. Thank you very much for bringing this matter to my attention. I will adopt a different attitude in my next round of meetings and visits to the farmers."

Stop the role-play

Discuss the attitude of Fati towards the farmers by considering these questions.

*Discussion Questions:*
1. Did Fati know initially that she had a blind window?
2. How did Fati realize aspects of her blind window?
3. What would she do to open the blind window?
4. How important is it to request for feedback on our behaviour?
5. What is the advantage of accepting feedback?
6. How will opening the blind window influence the way Fati will work with farmers?

Summarize the major points from the discussion, write them on newsprint for display on the wall.

**The Hidden Window**

The Hidden Window is an aspect of the personality that is known to oneself but unknown to others. One is, therefore, hiding this "window" from others for several reasons. It may be because one is afraid that if that aspect is known, colleagues may not like it. It may also be that one does not want a certain type of relationship to develop between them and others. It may be due to selfishness or the human climate prevailing in the community.

For example, an extension agent may unknowingly instruct a farmer to apply the wrong fertilizer to vegetables. The vegetables then wither in the field but the farmer does not know the cause. Both the extension agent and the farmer are worried about the issue. The extension agent is worried because he/she knows the cause of the problem but is hiding it. The farmer is worried because he/she does not know the cause of the problem but suspects the application of the fertilizer. The earlier the extension agent discloses the cause of the problem, the sooner he/she will recover from the guilt of it to work happily again with the farmer.

It is commonly stated that sharing problems with people lightens the burden by at least one half. The hidden window is opened through sharing. One is relieved by sharing what is hidden from others. As it is popularly said, *"sorrows shared are halved and happiness shared are multiplied"*.

On a happier note, if an extension agent has innovative methods of producing healthy vegetables and shares the information with farmers, they will have good harvests and words of praise for the extension agent. Moreover, they may have other brilliant ideas to supplement or complement the methods to increase crop production.

The main issue here is that extension agents should not keep the hidden window closed. They should open it through sharing. Both the extension agent and the farmers will then work together, trust each other and develop together. All suspicion and rumours between them will be minimized. Farmers will also be encouraged to share their problems and achievements for the common good.

### *Exercise 6.3 The Hidden Window*

*Purpose:*
The purpose of this exercise is to reveal aspects of the hidden window.

*Materials Required:*
Newsprint, coloured markers or crayons.

*Background to the role-play:*
Mr. Tetteh, a village extension agent in the Dangme West District of Ghana has collected 100 bags of fertilizer for his farmers but diverts them for sale to agricultural suppliers in Togo in order to get rich quickly. Throughout the growing season, his farmers continue to complain bitterly about the acute shortage of fertilizer for their crops.

Although he has been involved in this practice for the past three years without any notice, or complaints from farmers; the shortage of fertilizer during the current year was really severe. Mr. Tetteh's conscience began to bother him very much as the complaints of the farmers intensified. After brooding over the matter for sometime, he could no longer conceal this behaviour so, he decides to confide in a friendly farmer, Mr. Zoro, to reveal his deeds. As soon as he opens up and tells it all to Mr. Zoro, he is relieved through the advice and consolation from Mr. Zoro.

*Procedure:*
Ask one participant to act the part of Mr. Tetteh and another act as Mr. Zoro

Mr. Tetteh and Zoro engage in a conversation during which the real cause of the fertilizer shortage in the community is revealed to Mr. Zoro. Ask the other participants to observe the expressions on the faces of both players throughout the conversation.

Stop the role-play

Discuss the behaviour of Mr. Tetteh and why he continued to hide this aspect of his behaviour from his farmers.

*Discussion Questions:*
1. Did the farmers suspect that their extension agent was responsible for the fertilizer shortage?
2. Why did Mr. Tetteh confess his action to Mr. Zoro?
3. What changes occurred in Mr. Tetteh's behaviour when he revealed what he was hiding to Mr. Zoro?
4. How can we get over the worry of hiding something?

List answers on the newsprint.

Participants should be careful about the type of friends they share their problems and issues with. The story of two hunter friends can illustrate this caution.
    There were two hunter friends who always went on hunting expeditions together. One day the elder hunter showed a place to his friend explaining that he buried someone whom he mistook for game and killed. The friend in a haste then went to report the case to the chief. The chief

beat the gong-gong that midnight for the asafo company to go and dig out the corpse.

When it was dug, it was discovered that this was the predator animal, which the asafo company could never trace to kill for a prize of ten million cedis from the chief. The elderly hunter won the prize while the younger one lost his dignity.

**The Dark Window**

The Dark Window refers to the aspects of the self that are unknown to an individual and to other people. Many people have talents that they never develop since that window remains closed all their lives. An extension agent may become a highly successful TV actor if the dark window is developed.

There are many ways through which individuals can discover the dark window in order to improve or change the course of their lives and that of their neighbours or communities altogether. For example:

1. Adopting someone as a role model and aspiring to reach the level of that person's achievements;
2. Reading inspirational or challenging documents and books like the Bible and the Koran or books on adventure;.
3. Taking calculated risks or adventures;
4. Analyzing one's dreams and visions for a new direction;
5. Undergoing further studies or research, for example for higher degrees;
6. Engaging in serious moments of reflection;
7. Setting development goals and aspiring to attain them.

When some of these ideas are put into practice, the dark window will be opened, at least slightly, for the benefit of the individual and the community. Some known leaders at local, national and international levels have become great because of adventure or other ways of opening their dark windows.

Mahatma Gandhi's dream in 1930 about ocean breakers crashing on the rocks of the seashore inspired him to start the great salt walk to protest against the obnoxious British salt tax in India. (Everybody in India was expected to buy salt from the British at high prices). Many people joined the march spontaneously. The salt march which began on March 12, 1930 covering a distance of 241 miles became a symbol and a start of the fight for Indian independence which was won 17 years later. The salt march soon became world news and made Gandhi a national hero. As a result, the world's attention was drawn to the oppression of British colonial rule (Hope and Timmel, 1984).

### Exercise 6.4 Discovering the Dark Window

In the following example, we try to show that opening the dark window reveals the hidden talents of people.

An agricultural project manager, Mr. Gyedu, decides to venture into gospel music composition and production. He is inspired by how gospel singers are making it and are proclaiming the gospel of Jesus Christ and at the same time popularizing themselves and becoming rich. He thinks seriously about his vision for this venture. He finally starts this adventure while at the same time continuing with his agricultural project. Soon, he becomes a celebrated gospel singer, with many recordings to his credit. He is now able to supplement his income from the royalty proceeds from his cassettes. Had he not ventured to open his dark window, he could never have known that he had such a talent to develop into a well-known gospel musician.

*Discussion Questions:*
1. Did Mr. Gyedu ever realize that he was capable of developing into a famous gospel musician?
2. How did he discover that he had this talent?

List on the newsprint, some methods that we can adopt to discover our dark windows.

*Guidelines to Trainers and Facilitators:*
Note that these role-plays can be processed, discussed and evaluated in various ways. Each play can be discussed and evaluated separately at the end of the scene. Two or more role-plays can be evaluated together.

The most important outcome of the discussion and evaluation of each role-play should be lessons learned about human behaviour patterns and how these role-play examples can be used for our personal development and relationships with other people in our workplaces and communities.

CHAPTER 7

# Effective Teamwork

**The objectives of this chapter are to:**

1. explain that the success of an organization depends on the performance of the leader and team members;
2. show the importance of team performance review in the workplace;
3. identify the responsibilities of leadership.

**A Team**

A team is a purposeful and cohesive group with a sense of identity and clearly defined aims. Teams comprise persons having common interests in a specified task, and willing to take responsibilities and work together. Teamwork is very important because it is not always possible to accomplish certain tasks when individuals work independently. Teamwork is the driving force of any organization or association to survive and realize its objectives. Farming is a complex and difficult undertaking and therefore, demands that extension agents work together with farmers in order to achieve success. Here the idea of synergism comes into play: This is the condition where the combined effort of individual parts is greater than the sum of individual parts put together.

**Effective Teamwork**

It is very important for farmers and facilitators to maintain a strong sense of team spirit in Farmer Field Schools. Team spirit unites members of a team to successfully work together to avoid unhealthy competition and achieve their common goals.

Ensuring a coherent and effective team requires that we pay attention to:

1. The aims and objectives of the team;
2. Recommended working methods and procedures of the group;

3. Appropriate leadership styles;
4. Cordial relationships within the team;
5. Reviewing team performance and individual development.

**Aims and Objectives of Teams**

All teams exist for a purpose, but the way many teams work sometimes suggests that the team's purpose is unclear. The facilitator should ensure that the objectives of the teams are clear to all team members.

Objectives of teamwork should be SMART, where

S denotes specific objective
M — measurable objective
A — achievable objective
R — realistic objective
T — time bound objective

Each individual should clearly understand the aims and objectives of the team. Clear aims and objectives, particularly those arrived at in a democratic way lead to:

1. Greater motivation,
2. Greater creativity and initiative,
3. Less conflict,
4. Fewer demands on the leadership and
5. Better use of time and energy.

**Decision Making Procedures**

To obtain maximum commitment from team members, it is usually best to adopt the democratic method of decision making. People who have been involved in making decisions, are much more likely to accept and act on the decision made because they feel a sense of ownership of the decision.

Team leaders have to decide on the extent to which members take part in decision making. Each individual in the team should understand how he/she could contribute towards the attainment of the goals of the team and also respect the potential contribution of other individuals in the team.

*Effective Teamwork* 65

**Relationships within the Team**

To achieve maximum result, a team should develop good working relationships between members. Members can talk openly about disagreements and problems without fear of being attacked, ridiculed or punished in some way. A good team is built on the foundations of good communication, rapport, support, trust, co-operation, discussion, consensus and openness.

The principle of teamwork can be illustrated with a bicycle in motion. The front wheel and the steering (Fig. 7.1) can be regarded as performing the leadership function. The light shows the way in the dark. This is an aspect of the role of leadership. A leader must show the way, encourage full participation, identify problems ahead and solve them with the people.

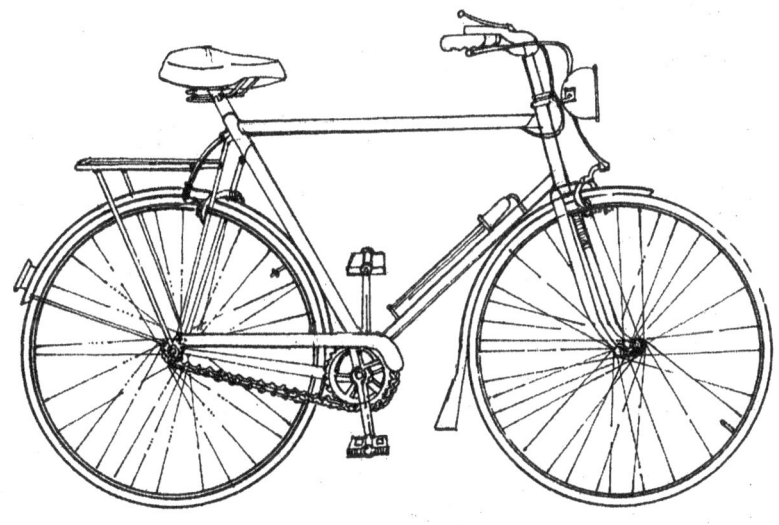

**Fig. 7.1:** *A bicycle*

In each of the bicycle wheels, there are many spokes that are attached to the wheel rim. Both wheels are connected by the bicycle chain and all parts work together to make the bicycle move. The wheels of the bicycle can be considered as the group of farmers who are united and work together with the leader to make decisions together in order to achieve the objectives of the team.

### Exercise 7.1 Teamwork Illustrated by the Parts of A Bicycle

*Purpose:*
To demonstrate the importance of successful teamwork.

*Materials Required:*
Drawing of a bicycle with the parts labeled, newsprint, coloured markers.

*Procedure:*
Participants should arrange themselves in their working groups.

Put up the drawing of the bicycle in front of the group or distribute copies of the bicycle drawing to working groups.

Distribute newsprint to the working groups.

Ask participants to list the parts of the bicycle.

Participants should identify the parts that they consider most important.

Hold a general discussion, noting that there are no right or wrong answers in this activity.

*Discussion Questions:*
1. Can the bicycle run efficiently without any of the parts?
2. Which parts of the bicycle are the most important and why?
3. Do the parts of the bicycle work together like members of a team?

### Exercise 7.2 The Water Brigade — Demonstrating the Importance of Planning for Successful Teamwork

*Purpose:*
This exercise provides the opportunity for participants to understand the importance of effective planning for successful teamwork.

*Materials Required:*
Two buckets, two large basins, and enough water to fill the buckets.

*Procedure:*
Participants should divide themselves into two groups.

Each group should stand in a line facing the other.

Place a bucket full of water at the front end of each line.

Place an empty basin at the other end of each line.

Explain to participants that, using only their palms, each team should transfer the water from the bucket into the basin at the end of each line. To do this, the person at the front of the line will take water from the bucket in his/her palms and pass it on to the next person who passes it on to the next person and so on until the last person empties his/her hand into the basin at the end of the line.

Give the start signal and supervise the exercise.

Compare the quantities of water transferred into the basins by the teams.

At the end of this exercise, the group should determine which team finished first and which team transferred more water from the bucket into the basins.

Hold a general discussion.

*Discussion Questions:*
For the group that finished first?
  1. Why did the team finish first?
  2. What plan or strategy did this group adopt to make them successful?
  3. Did this group transfer more water than the other group? Why?

For the group that finished second:
  1. Why did this group fail to finish first?
  2. What plans did the group make?
  3. Why did their plan not lead them to be successful?

List the major factors that determine the success of a team.

## Reviewing Team Performance

A team needs to review its performance regularly when the facilitator creates favourable environments for these reviews.
Some of the benefits of regular review include:

1. Enhancement of efficient and democratic decision making;
2. Reduction in the numbers and frequency of crisis;
3. Identification of future needs and effective planning for the future;
4. Commitment of team members to team objectives;
5. Improvement in leadership as team leaders learn from experience and feedback;
6. Identification of individual weakness and strengths and
7. Promotion of individual development.

## Relationships with Outsiders

However effective your team is, it cannot survive on its own. It is part of a large system, which includes other teams and individuals, departments, sponsors, customers, traditional authorities and other interest groups. Members should ask for help, opinions, ideas from other teams, their leaders and experts. The key player in a team building process is the leader.
The next chapter considers Leadership.

CHAPTER 8

# Leadership

**Objectives of this chapter are:**

1. to discuss the concept of leadership;
2. to illustrate the four major leadership styles;
3. to use animal codes to explain human behaviour in groups.

**Introduction**

We start to discuss the issue of leadership with the definition of the concept. Leadership can be defined in many ways. It may be referred to as "a position of influencing others to act", "getting people to do things willingly", "a position of authority" "the way of managing people and resources", "the art of organizing people" (Mullins 1993). According to the Longman's *Dictionary of Contemporary English,* a leader is a person who moves, directs a group or an organization or a person who is ahead of others. Leadership is a status with very many responsibilities attached to it. It also prescribes certain leadership behaviours which according to House, include the following:

1. Directive leadership:— letting subordinates know and perform their tasks.
2. Supportive leadership:— displaying love and concern for others.
3. Participative leadership:— consulting with others before making decisions.
4. Achievement-oriented decision leadership:— setting goals for the work for subordinates and encouraging them to perform.

It is important that extension agents understand the roles and responsibilities of a leader and leadership styles to adopt at various levels and situations.
Leadership can be found at any level of society or group. Leadership is knowing the way and leading people to perform creditably. It relates to motivation, delegation of power, team building and interpersonal relations.

Leadership has the moral and professional ability to visualize strategies for the success of organizations. It inspires subordinates and colleagues and develops teamwork to enhance effective performance. It promotes individual and organizational performance.

Leaders may be appointed or elected. Sometimes, people may usurp leadership naturally because of charisma, experience, knowledge or training, wealth, education, long service, heredity, and hard work. Some people usurp leadership with brute force. Some leaders behave as Moses did among the Israelites in the Bible. They assume that their *modus operandi* is the best as Moses did when he killed the Egyptian to save his own countryman. To his surprise, his action was condemned by his compatriots. Moses had to run away to save his own life. Leaders need to study and know their followers and situations so that they can live and work together as a team. Whatever the source of leadership, leaders can improve their performance through training.

**Leadership Styles**

Leadership styles refer to how leadership functions are performed and how the leader relates to subordinates. There are many dimensions to leadership styles. This is because there are many leadership styles as there are leaders and leadership is exercised in different environments and situations.

Four major leadership styles are identified by Mullins (1993). These are the *autocratic*, *laissez-faire*, the *bureaucratic* and *democratic* styles. In any case, however, the situational approach to leadership must be adopted. There is no single leadership style suitable for all occasions. Leadership styles are adopted according to the demands of the situation.

Douglas McGregor's X and Y theories categorize leadership into two major styles. These are the *task-centred* and *people-centred* leadership styles. McGregor's theory X explains that human beings by nature are lazy, avoid responsibility, have low ambition and inherently dislike working. In these contexts, people must be forced, directed, controlled and punished to keep them busy. The task-centred approach to leadership is employed combining both the bureaucratic and autocratic (authoritarian/dictatorial) styles. Leadership in this sense enforces institutional structures and regulations and ensures compliance (McGregor 1960).

By theory Y, McGregor assumes that humans have the natural

tendency to work, accept responsibility and can develop and utilize potentials to contribute to personal progress and society. In this case, leadership is people-centred. Theory Y then supports the use of the democratic and laissez-faire leadership styles. Emphasis is placed on the building of cordial relationships with subordinates. The leader listens, encourages and affirms subordinates in whatever they say or do. They are taken into partnership in the decision making process and allowed to take initiatives.

Leadership styles must be modified according to the situation, the maturity of people, the goals of the organization and the general environment.

**Four Leadership Styles**

*Autocratic Leadership Style (Fig. 8.1)*
The autocratic or authoritarian style of leadership is bossy; the leader assumes knowledge of everything. For example, an autocratic extension agent assumes that only he/she knows all what the farmer needs. Such an extension agent interacts with the farmer using a dictatorial approach, always giving instructions to the farmer. No questions are entertained from the farmer who is continuously loaded with instructions. If the crops fail, it is because the farmer has refused to listen to the "expert advice" of the extension agent. He/she finds faults with the farmer most of the time, looking down on and instilling fear in the farmer who often becomes confused. Any time the autocratic leader approaches the farm, the farmer becomes worried and sad. This kind of leader is the only person who exercises power at the policy, administrative, decision making, reward and punishment levels.

The autocratic leader does not take advice and believes that his/her line of action and thinking are always the best. Such leaders do not tolerate any ideas that do not originate from them. They are annoyed if approached by subordinates. As far as they are concerned, problems should be solved at the lower level. When they are present, everybody seems to be sad although each person pretends to be working hard. When danger is imminent no one can approach him/her for direction or give advice until the community disintegrates. It is pathetic to live under such a type of leadership.

72  Non-Formal Education for IPPM

*Fig. 8.1*: *The autocratic leadership style claims to know more than everybody else does.*

*The Laissez-faire leadership style (Fig. 8.2)*
The laissez-faire type of leader allows some measure of freedom to subordinates to work on their own. Each feels committed to work. This style is viable among professionals who can work with little supervision. The leader co-ordinates the activities of all. But in extension work among illiterate communities, this leadership style degenerates into non-style leadership and may be best labelled as abdication of responsibility leadership.

This is the carefree type of leadership. An extension agent with the characteristics of a laissez-faire leader is lazy and does not take his/her work seriously. Problem areas are often avoided. Decision making is in the hands of workers. He/she does not frequent the cropping fields, is not worried about advising farmers to plant although the rains may have started. Efforts are not made to negotiate for inputs for the farmers and does not encourage them to obtain the inputs on time. Yet he/she collects his salary before other extension agents are aware that salaries have been paid.

The guiding motto for this kind of leader is *"country broke o, country no broke o, we dey inside"*. Such behaviour and attitude to work do not

Leadership 73

inspire farmers who, therefore, continue to practise inefficient methods of agricultural production.

Even when the laissez-faire extension agent finally takes the trouble to introduce any new input or modern method of farming, there is no follow-up to find out if the message has been well understood and adopted by the farmers. At meetings, a laissez-faire leader accepts divergent views proposed by members.

*Fig. 8.2: Laissez-faire leadership style*

*The Bureaucratic Leadership Style (Fig. 8.3)*
This type of leadership complies rigidly with rules, directives, correspondence and regulations from his/her superiors. For example, an extension agent that behaves like a bureaucratic leader advises farmers to keep strictly to laid-down planting schedules irrespective of differences in the rainfall patterns in different years. Whenever anything goes wrong, such an extension agent immediately blames the regulations or the bosses. He/she has no empathy for people, and satisfies the employers. The official closing time is

strictly respected and, therefore, no extra time can be spent to attend to farmer's problems however pressing they may be. The bureaucratic leader behaves officiously and hides behind "General Orders".

The bureaucratic extension agent who convenes a meeting is always punctual. He/she condemns all those who come to the meeting late because everything must be done as laid down by rules. He/she never listens to the excuses given by farmers. Continuous reference is made to authority or correspondence from authority to support all actions. This leadership style does not have a human face.

*Fig. 8.3: The Bureaucratic leadership style.*

*The Democratic and Participatory Leadership Style (Fig. 8.4)*
The democratic and participatory type of leadership relates well with people and involves them in decision making at all times. In this way, there is always consensus and everybody concerned feels a strong sense of ownership of decisions.

For example, the democratic and participatory type of extension agent spends considerable time discussing issues with farmers and making collective decisions. He/she knows farmers by names, visits them in their homes and is concerned about their welfare. He/she becomes an adviser

*Leadership* 75

to farmers not only on farming issues but also on their general welfare.

The democratic extension agent relates cordially with farmers, sharing their problems and aspirations with them. He/she takes them into close partnership, working on the field with them, organizing farmer training workshops and ensures that they practise what they learnt at the workshops. During workshops, a very friendly environment is created for everybody to participate in the discussions. Quiet participants are invited to speak up and congratulated for their contributions. He/she cleverly points out what has gone wrong and always treats farmers as colleagues.

*Fig. 8.4: Democratic leadership style.*

*Exercise 8.1 Types of Leaders*

*Purpose:*
The purpose of these role-plays is to illustrate the major characteristics of the different leadership styles and to highlight some virtues in leadership.

*Materials Required:*
Newsprint, coloured markers, sellotape.

*Procedure:*
Set up four groups: A, B, C, and D each of six participants, to act as farmers attending a meeting or interacting with an agricultural extension agent.

Ask four participants to volunteer to act as extension agents.

Brief them, together with the individual farmer groups, separately, to act as described in Play A, Play B, Play C, and Play D.

Each play should be performed, one after the other, in front of the entire group.

Allow 15 minutes for each play, which should be stopped when the message has been clearly illustrated.

After all the plays have been performed, discuss the characteristics of the leaders by comparing the behaviour patterns of the extension agents in the role-plays.

Use the discussion questions given at the end of this section as a guide in the general discussion of the plays in order to characterize the leadership styles.

*Play A*
At a meeting of farmers, the extension agent, Mr. Francis Tagoe who summoned the meeting is absent. All farmers already know about Mr. Tagoe's autocratic behaviour but they are afraid to confront him. Farmers are disappointed and begin to discuss and complain about his unacceptable behaviour. As they discuss this issue among themselves, Mr. Tagoe arrives. He shouts at the farmers to stand up to welcome him as a sign of respect. As the meeting progresses he continues to shout at any suggestions farmers make and refuses to answer any questions from them. Rather, Mr. Tagoe constantly reminds farmers that he was trained overseas and that they must listen to and obey instructions from him at all times. These dictatorial tendencies make farmers quiet, submissive and sad.

## Play B
Weija Tomato Farmers Association has gathered together for a meeting organized by Ms Abena Lamptey, the extension agent. The farmers wait anxiously and are worried because the convener is nowhere to be seen and no message was sent to provide any information; finally the extension agent turns up after two hours. On arrival, Ms Abena cannot direct affairs at the meeting. She is absent-minded and occasionally interrupts the meeting with irrelevant issues like the high prices of jewelry, and women's social problems in Ghana. Farmers are disappointed and start discussing their personal domestic problems.

## Play C
The scene is a farm where the farmers are interacting with their extension agent, Mr. Hassan Abdullahi who is carrying a cropping calendar for maize. The previous year, the rains were very late and farmers lost their maize crop because they strictly followed the cropping calendar given to them by Mr. Hassan. This year, the rains are late again yet Mr. Hassan is insisting that farmers must adopt the same cropping calendar because these are the instructions from the headquarters of the Department of Agricultural Extension Services in Accra. Farmers argue vehemently with the extension agent who bluntly refuses to modify the cropping calendar according to the prevailing rainfall pattern. He continues to show them the calendar as the source of information. The farmers become fed up with Mr. Hassan and leave the site one after the other.

## Play D
The extension agent, Ms Patience Bruce-Doe, calls farmers to a meeting and is very punctual at the meeting. As she arrives, she greets them and enquires about the welfare of their families. Patience involves all the farmers in the discussion and prompts the silent ones to speak up and make suggestions. She identifies herself with their problems and through teamwork, practical solutions are reached to solve problems. Farmers show signs of satisfaction and happiness because of the friendly and participatory nature of the meeting and the attitude of Patience to them.

*Discussion Questions:*
1. What happened in plays A, B, C, and D? Describe exactly what you saw in each play.

2. How would you describe the behaviours of the extension agents, Mr. Francis Tagoe, Ms Abena Lamptey, Mr. Hassan Abdulahi and Ms Patience Bruce-Doe?
3. List the major features of the behaviours of the extension agents in play A, B, C and D.
4. What are the major differences between the behaviours of the extension agents?
5. How did the behaviour of the extension agents affect the farmers?
6. In which situation do you think that each type of leadership style is appropriate?
7. How does the leadership style contribute to achieving group objectives?
8. Which of these leadership styles is appropriate for working with farmers or with colleagues?
9. Summarize the major outcome of these discussions on newsprint and paste them on the wall of the classroom. Lessons can be drawn on good leadership styles but participants should be reminded that, leadership is basically situational. There is no one leadership style appropriate for all situations. The appropriate contingency model must be adopted. Under normal circumstance, however, leadership must be democratic.

## Human Behaviour within Groups

In the first section of this chapter, we discussed leadership roles that are important for group cohesion and productive work. Members of groups are always excited about their success as a group. However, people generally behave differently because of their background and interest, particularly during times of stress. Sometimes, people realize, rather painfully, that their behaviour has not been very helpful to the group. We can analyze human group behaviour patterns by using animal codes that help us to examine the positive and negative aspects of our behaviour within a group. Leaders need to know about different behaviour patterns in the group so that they can adopt the appropriate leadership styles to enhance teamwork.

But before then, here is a joke to brighten the session.

# Leadership

> **BOX 8.1**
>
> A gentleman gave a lift to a lady in his car. On the way, the gentleman kept on looking at the lady. The lady suspected that the gentleman wanted to propose something but could not do so. When the lady alighted from the car, she left a note in the car and pleaded with the gentleman to read it on arrival at home. The gentleman quickly negotiated a curve, stopped and read the note which stated: "Read Mathew 7;7".
>
> He recollected the text in the Bible, which states "Ask and ye shall receive; knock and it shall be opened onto you, seek and ye shall find". The man immediately made a sharp turn to search for the lady. But the lady was nowhere to be found. He missed her desperately.
>
> Lesson: Leadership needs initiative.

**Exercise 8.2 Using Animal Codes to Analyze Human Behaviour Patterns**

*Background and Purpose:*
This exercise on animal codes is useful for explaining how some people and even leaders behave. The exercise also helps us to understand ourselves in order to improve upon our working and human relationships.

Thus, the exercise will provide the opportunity for personal reflection and growth and group sharing of experiences leading to the improvement of community life and work.

This exercise is based on the Enneagram, a Greek word meaning the "diagram of nine" which portrays the nine typical patterns of compulsive behaviour among people. Each person, like the animal illustrated, exhibits a dominant or compulsive behaviour. Animals very clearly show some kinds of human behaviour patterns. This exercise is adapted from Hope and Timmel (1984) and the text from Dorr (1990). The nine animals discussed are the ant, the spider, the dog, the cat and the owl. Others are the deer, the monkey, the lion and the hippopotamus. In addition to the nine character types, some other animal characters have been discussed.

*Materials Required:*
Large pictures of the animals specified in the texts. (Fig. 8.5–8.22).

*Procedure:*
Mention the name of each animal. Ask members of the group to volunteer to describe the common characteristics or traits of each animal.

After each description, the picture of the animal is presented for all to see and discuss the particular behaviour described. Participants are invited to share ideas on the picture. The main behavioural characteristics of animals are raised. Unco-operative attitudes are noted. Emphasis is placed on the good qualities of each of them and the dominant compulsive behaviour is highlighted. The dominant or compulsive behaviour of each animal is noted and written below each animal in the picture.

After the discussion of each picture, ask participants to examine themselves and to find out whether the illustration has given them some insight into their own behaviour or that of others in the community.

The dominant character types in the group can also be discussed. Participants should suggest ways of improving their lives in response to the points discussed. List these on newsprint for further group discussion.

*The Compulsive Worker*
This is the workaholic who despite all odds continues to work hard all the time, finds a lot of pleasure in working. The type of person who feels that the only way to be recognized is to work hard. Is identified by hard work and is an achiever and tries as much as possible to avoid failure.

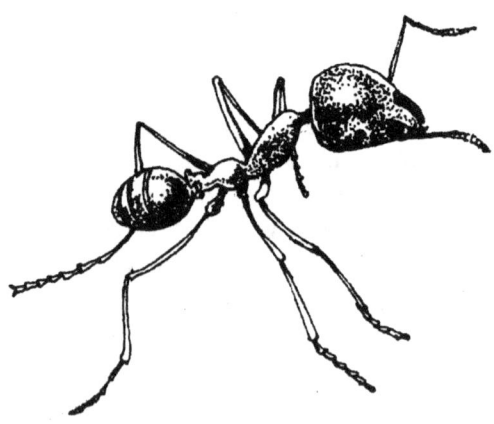

***Fig. 8.5:** The ant (The compulsive worker)*

The compulsive worker is very committed to the work and when in charge of a project may turn out to be a slave driver. He/she is honest and caring. He/she does the work which others refuse to do and is thus overworked. Has little time to rest.

## The Compulsive Perfectionist

Like the spider (Fig. 8.6), which makes and maintains an elaborate and beautiful web, the compulsive perfectionist is the leader who wants "to get things right". The perfectionist is not content with the way things are. He/she has developed the culture of maintenance and wants everything to be done perfectly. He/she is anxious to correct people and help them improve upon their lives and conditions. He/she is generally concerned with the improvement of everything in the environment. Is punctual and disappointed when others are late. His/her compulsion to help improve situations sometimes results in conflict. Becomes angry at poor conditions and people who do not change for the better, and deals drastically with lazy people.

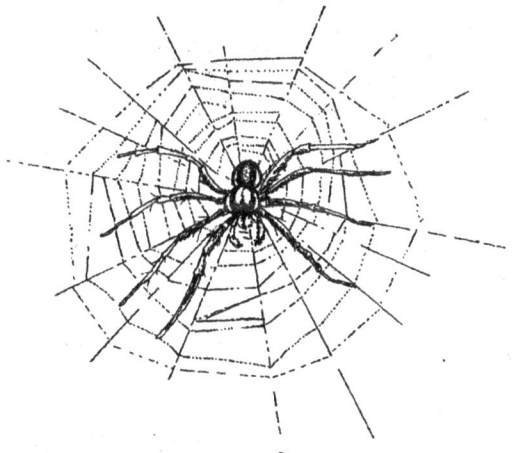

*Fig. 8.6: The spider (The compulsive perfectionist)*

## The Compulsive Helper

The compulsive helper is a person, like the dog (Fig. 8.7), who is always anxiously protecting others, caring for them and looking after them at his/her own expense. While others are at rest or enjoying themselves, he/she is alert protecting their interest. Always anxious to help but is also expecting a reward. Needs to be appreciated and loved although does not show that he/she needs attention or demands it. However, because the compulsive helper does not ask for rewards but continues to serve others, the help is taken for granted. He/she is often forgotten or not adequately rewarded, and can develop grievance against his employer or against those who enjoy the fruits of his/her hard labour. This situation can be likened to the biblical prodigal son and his brother. His brother dutifully served his father over a period but was not rewarded. When the prodigal son returned, he was given a red carpet treatment. His brother (the dog) who had all this time been serving religiously now revolted.

*Fig. 8.7:* The dog (The compulsive helper).

*The Compulsively Special Person*
The compulsively special person, like the cat (Fig. 8.8) is the type of person who feels very different and, therefore, special. Is sensitive to beauty and is conscious about personal presence and appearance. Always looking for and demanding sympathy and when a sympathetic listener has time, will go on narrating all about himself. He/she is jealous of others who make the same demands. He/she believes that he/she has been contributing a lot towards the development process but has been suffering.

*Fig. 8.8: The cat (The compulsively "special person")*

## The Compulsive Observer

The compulsive observer, like the tortoise (Fig. 8.9) looks very solemn and seems to be very knowledgeable and wise. More interested in thinking rather than action. Believes that the only way out of an insecure environment is to think about and observe what is happening around him/her. But more often than not, such assumptions are incorrect. Does not contribute ideas or money but is rather selfish. Does not give time or energy to group development, needs time to think and accumulate knowledge, hates to be

*Fig. 8.9: The tortoise (The compulsive observer)*

disturbed in any way and likes to maintain privacy. Silent and inactive at meetings and at work respectively, most of the time absent-minded even in meetings or at work. Does not want to talk of work so as to make mistakes and be criticized.

*The Compulsive Loyal Person*

The deer (Fig. 8.10) is a very submissive animal. A person who is likened to the deer harbours feelings of insecurity, becomes part of the group, which is believed to be secure and identifies interests with that of the group; moves along with the group. Complies with the laws of the society and will not like to appear as a deviant. The greatest value is, therefore, respect for the laws and tradition of the group. Has great respect for authority. Anxiety for security is the driving force for this compulsive behaviour. Does not take decisions alone and does not want to offend anybody. Prepared to undertake any challenge and tasks in order to conform with norms and regulations. Becomes authoritative when put in power and does not want this position to be threatened. If this happens, becomes very vicious. But most of the time, the compulsive loyal person is friendly to all and avoids confrontation.

***Fig. 8.10:*** *The deer (The compulsive loyal person)*

*The Compulsive Optimist*
Like the monkey (Fig. 8.11), the compulsive optimist avoids pain and is interested in pleasure; jokes around often and has developed a large appetite for enjoyment. When others are seriously discussing important issues or are at work; interrupts them with trivial issues. He/she goes to the extent of disrupting the group from concentrating on any serious business, entertains the group with exciting stories and any kind of issues that may keep people's spirit high. He/she does not take anything serious but plays down or ignores any problems and failures. If problems cannot however be ignored, the optimist takes consolation that it is the will of God and that better days are ahead, and therefore, always living on hope. Lacks self-control even in the face of very serious issues and can be in trouble for interference.

*Fig. 8.11:* *The monkey (The compulsive optimist)*

*The Compulsive Aggressive Person*
Like the lion, (Fig. 8.12) the compulsive aggressive person is interested in power and, therefore, looks for it wherever it can be found in the community or in the group; is an assertive type of leader who can inspire others to work. Becomes a bully when others do not agree with him/her. Is aggressive,

courageous and forceful. The aggressive nature reflects not only in actions but also in speech. Does not admit guilt but is always on the alert to point out the mistakes of others. Always ready to defend rights and those of the people very close to him/her. Tramples on the rights of others. Demands respect and crushes down on competitors or those whose presence seem to threaten his/her position.

*Fig. 8.12:* The Lion (The compulsive aggressive person)

*The Compulsive Peacemaker*
The compulsive peacemaker like the Hippo (Fig. 8.13) is the personality type that places high value on peace keeping. Does not want to be disturbed and would not also disturb others, therefore does not take initiatives or decisions but allows others to do so. Becomes an active participant in activities. Avoids conflict with others, is patient, tolerant, unassuming and can get along with any character type. Accepts people as they are without making any effort to change them. Is all loving, but not inspiring. Like the hippopotamus however, when survival is threatened, reacts violently.

Leadership 87

**Fig. 8.13:** *The hippopotamus (The compulsive peacekeeper)*

*The Stubborn Person*
The stubborn person, like the donkey (Fig. 8.14) will not change from an original point of view. Such behaviour tends to slow down progress within a group.

**Fig. 8.14:** *The donkey (Stubborn)*

### The Ostrich-type of Person

The ostrich (Fig. 8.15) buries its head in the sand while exposing the rest of the body and cannot face reality to admit that there is a problem. This kind of behaviour thrives on self-deceit, believing that there is no problem whereas he/she lives in a world full of problems.

***Fig. 8.15:*** *The ostrich (Pretends there is no problem)*

### The Rabbit-type of Person

The rabbit (Fig. 8.16) a smart and intelligent animal, runs away as soon as there is conflict, tension or an unpleasant situation. This is also described as flight behaviour. The rabbit type of a person, tends to dodge problems and quickly switches to another topic to avoid unpleasantness.

***Fig. 8.16:*** *The rabbit (Dodges problems)*

## Leadership

### The Obstructionist
The obstructionist behaviour, like the elephant (Fig. 8.17) blocks the way and stubbornly prevents the group from achieving desired goals. This is described as obstructionist behaviour. On a positive note; the elephant sometimes clears the way for its followers and prepares the environment for them to enjoy the facilities available.

*Fig. 8.17:*     *The elephant (obstructionist)*

### The Arrogant Person
Arrogant people behave like the proud peacock (Fig. 8.18) always showing off and competing for attention.

*Fig. 8.18: The peacock (proud)*

## The Dangerous Person

The snake (Fig. 8.19) always hides in the grass and strikes unexpectedly. Persons with this pattern of behaviour are usually dangerous and difficult to deal with in a group.

*Fig. 8.19:*     *The snake (dangerous)*

*Leadership* 91

*Superiority*
Some people adopt the behaviour of pride and superiority; just like the giraffe (Fig. 8.20) which depicts superiority, looking down on others.

***Fig. 8.20:***   *The giraffe (harbouring superiority complex)*

*The Compulsively Timid Person*
The compulsively timid person behaves like the mouse (Fig. 8.21). Too timid to speak up on any issue and does not expose its presence. This behaviour is destructive in a group.

*Fig. 8.21: The mouse (very timid)*

### The Chameleon Type of Person

The chameleon (Fig. 8.22) always changes colour according to the situation, saying one thing to one group and something completely different on the same subject to another group. People of this nature cannot be trusted since they change to please those in authority. Community workers who change according to the environment can work with all kinds of groups effectively. But they must be principled and should not only change to please people.

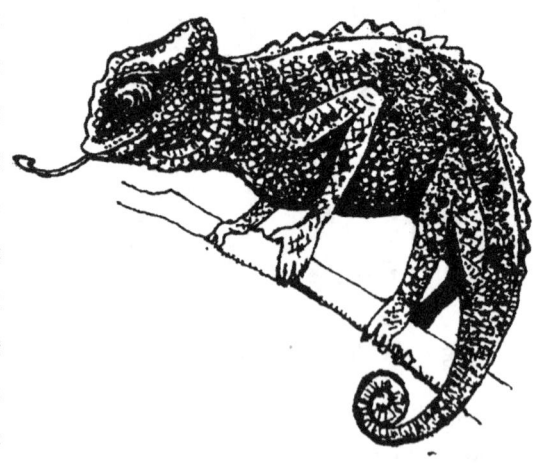

*Fig. 8.22: The chameleon (unstable and deceptive)*

**BOX 8.2**

## Some Guidelines for Effective Leadership

### Be A Leader Not A Boss

The Boss Drives His Men
The Leader Inspires Them.

The Boss Depends on Authority
The Leader Depends on Goodwill.

The Boss Evokes Fear
The Leader Radiates Love.

The Boss Says "I"
The Leader Says "We"

The Boss Shows Who Is Wrong
The Leader Shows What is Wrong.

The Boss Knows How It Is Done
The Leader Shows How It Is Done.

The Boss Demands Respect
The Leader Commands Respect.

## So be a Leader not a Boss

# Decision Making in Integrated Production and Pest Management: Agro-ecosystem Analysis

The objectives of this chapter are:

1. to explain the concept of group decision making and consensus building;
2. to demonstrate the application of group decision making in integrated production and pest management.

## Introduction

The most important activity of integrated production and pest management (IPPM) training in Farmer Field Schools is agro-ecosystem analysis, commonly abbreviated as AESA. AESA is a decision making tool which enables farmers to make informed production and pest management decisions based on careful scientific observations of ecological relationships in their fields, between crops and various abiotic and biotic components in the cropping environment.

To conduct AESA, farmers work in small groups visiting their fields regularly each week to make observations that are recorded and represented graphically, analyzed, and discussed together to reach a consensus on appropriate interventions to improve crop growth and performance. Through an understanding of the interactions between components of the crop ecosystem, farmers can maximize profits without destroying the ecosystem while at the same time conserving natural bio-diversity. High levels of group dynamics and interpersonal relationships are essential for the success of agro-ecosystem analysis. Various adult learning processes take place during the process of agro-ecosystem analysis to facilitate discussion and decision-making.

We have, therefore, included AESA exercise in this Field Guide in order to emphasize aspects of learning processes and interpersonal relationships that are vital for the success of this important activity. AESA forms a sound scientific basis of decision making in IPPM. Trainers and facilitators need thorough understanding of the principles and learning processes of AESA in order that they can confidently assist farmers to conduct this activity successfully.

## Exercise 9.1 The Process of Decision Making in IPPM

*Purpose:*
We can identify two major aspects of agro-ecosystem analysis. A technical aspect concerns the agronomic observations of the plants, the biotic and abiotic factors that influence plant growth and are used for analysis of the field situation. A second aspect is social, and concerns the learning processes and interpersonal relationships which promote group dynamics, leadership skills and consensus building.

The purpose of this exercise is to examine and pay attention to the social aspects of agro-ecosystem analysis and to explore how these components can be strengthened for successful decision making in integrated production and pest management and community development.

*Materials Required:*
Large sheets of newsprint, pencils, erasers, coloured crayons and coloured felt markers, rulers, field notebooks.

*Procedure:*
This exercise is designed for trainers and facilitators. The AESA process is described in greater detail in the *IPM Extension Guide Number* 1 to which reference should be made. Now proceed as follows:

1. Divide participants into small working groups of five persons each. Take participants to the field. Walk across the field and choose 10 sample plants in one plot. For each plant, participants should make the following observations and records.

2. *Plants:* Measure the size of the plant, the height, count the number of leaves (yellow or green leaves), flowers or fruits.

3. *Insects:* Carefully observe, count and record the numbers of insects and other arthropods seen on different parts of the plants.

4. *Beneficials and other Natural Enemies*: Look carefully at the plants,

count and record the numbers of spiders, mantids, other natural enemies and beneficial arthropods seen. Look for and count the numbers of any parasitized insect eggs and larvae seen.

5. *Diseases:* Look carefully at the leaves, stems and fruits of the plants. Observe any leaf colouration caused by plant diseases. Count and record the numbers of leaves, stems or fruits affected. Estimate the percentage of leaf area infected by diseases.

6. *Weeds*: Count the different types of weeds seen in the plot.

7. *Rats:* Look carefully for plants damaged by rats and record the numbers of plants affected.

8. *Soil and Water:* Examine the soil and determine the state of the soil, whether the soil is dry or moist.

9. *Weather:* Record the weather condition, which may be sunny, cloudy, rainy, hot or cool.

10. After these observations, participants should now seek a shady place under a tree. Each group should sit together in a circle, with pencils, coloured crayons and felt markers, data from the field, and the AESA record sheets of the previous week.

11. **Processing AESA data**: Participants should now make drawings of their field records on the large newsprint with all members participating fully and paying attention to the following.

    i. Draw the plant with the correct numbers of stems and leaves. Write the plant height, number of green leaves, number of yellow leaves. If the plant is healthy, colour it green; if the plant is diseased colour the plant accordingly, Yellow leaves should be coloured yellow.

    ii. Draw the different kinds of insect pests seen on the left hand side of the plant; write the numbers of each type of insects by the side of each drawing. Write the local names of each insect.

*Decision Making in IPPM* 97

iii. Draw the spiders and other natural enemies and beneficials seen on the right hand side of the plant; write the numbers and local names of each type seen next to each drawing.

iv. Draw the weather condition on top of the plant; for example if it is sunny, draw a bright sun and if it is rainy, draw rain showers falling on the plant. Draw clouds if it is cloudy.

v. Draw the kinds of weeds showing their approximate size in relation to the size of the plant. Write an estimate of the density of the weeds.

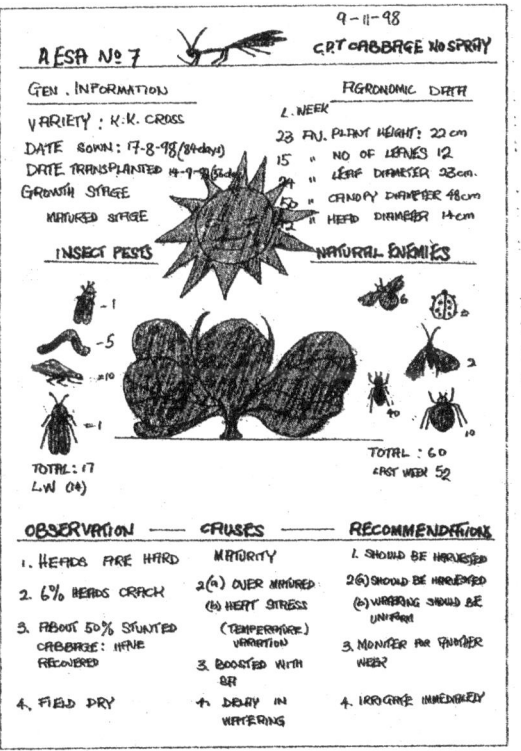

*Fig. 9.1 An example of an AESA record sheet.*

Source: Youdeowei, Anthony (2001) *Principles of Integrated Pest Management. Growing Healthy Crops.*

12. Each group should now discuss the field observations and arrive at a consensus on the priority factors influencing plant growth during the particular crop growth stage. At the bottom of the AESA sheet, under the drawing of the plant, they should write a summary of their observations, the possible causes and specific recommendations for action. (see Fig. 9.1)

13. Each group should select a member to make the presentation of their field observations, drawings, discussions and recommendations to the large group.

*Discussion Questions:*
After the group presentations and analysis of the technical aspects of the AESA exercise, participants should now address the following questions, which focus attention on the social aspects of agro-ecosystem analysis.

1. During fieldwork, how did participants interact with one another? How would they describe the levels of communication and dialogue between them?

2. Which of their senses did participants use when making observations in the field?

3. How did continuous exposure to the field and the crops there facilitate learning about the agro-ecosystem?

4. During processing of the AESA data, did all members participate? If not, what prevented some members from participating?

5. To what extent does graphical representation of field observations aid retention and learning? And does the use of different senses promote learning?

6. Who among the group members assumed leadership, or was appointed the group leader to direct the activity of the group? How did the person assume leadership? If the group chose a leader, what criteria were used to choose the leader?

7. How would they describe the group leader and style of leadership?

Did the leader encourage participation by all members of the group?

8. Group dynamics: to what extent was there a pleasant and friendly atmosphere in the group? How did the prevailing atmosphere influence the work of the group?

9. Did members of the groups share responsibilities effectively? If not why?

10. How was the learning process facilitated in the group setting through exploiting the individual strengths of members of the groups?

11. Decision making: How did the groups arrive at their decisions? Do all members of the group accept the group decisions and feel able to defend them?

12 How does the agro-ecosystem analysis exercise help to develop communication skills, dialogue, critical thinking, consensus building and confidence of people, especially farmers?

*Exercise 9.2 Role play on Decision Making: Selecting a Partner*

This role-play is taken from *Group Process and the Inductive Method* by Carmela D. Ortigas. It is best used in the training of Extension agents; it is not appropriate for farmers.

*Objective:*
The objectives of this role play are to
1. practise making individual and group decisions and consensus building.
2. discuss factors that determine individual and group decision making.

*Materials required:*
Flip Chart stand and papers
Coloured Markers

*Procedure:*
1. Divide the group into two smaller working groups; males and females.
2. The groups should sit in two separate areas.

3. Ask five volunteers from each group to sit in the inner circle, while the rest sit in an outer circle.
4. From the five female volunteers, the males should each select a partner.
5. From the five male volunteers, the females should each select a partner.
6. Each volunteer should now write down the reasons for choosing the respective person as a partner.
7. The persons in the inner circle then discuss together their individual decisions while those in the outer circle take notes of the processes that the inner individuals adopt in their selection and discussion.
8. On the basis of individual decisions, the entire group of women formulates a set of criteria for selecting male partners; they should write the major criteria on the flip chart.
9. The group of men also formulates a set of criteria for selecting female partners; they should write the major criteria on the flip chart.
10. The men and women should then compare the processes adopted for setting criteria for decision making.

*Discussion Questions*

At the end of this role play, conduct a group discussion with the entire group on the activity as a structured learning experience in decision making. Use the following questions to direct the group discussion

1. How did individuals arrive at their decisions? What were the steps adopted by the groups to arrive at their decision?
2. Did all members of the groups agree with the decisions?
3. What factors influenced individual and group decision making?
4. What are the differences and similarities between individual and group decision making? List these on a flip chart.
5. Discuss the general conclusion from this role play.

CHAPTER 10

# Brighten Your Training Sessions

**The objective of this chapter is:**

1. to provide items (ice-breakers) which could make adult training sessions interesting.

**Introduction**

Jokes, proverbs and short stories are useful energizers to enhance the adult learning process. They can be used to arouse and sustain participants' interest in the learning process. Jokes and proverbs help adults to think critically, learn about life and solve some problems. Some jokes and proverbs are given below to guide the trainer or facilitator to brighten up the training sessions by removing boredom to make the sessions interesting. These jokes and proverbs should be introduced periodically without any specific schedules. Most of these jokes are set in the Ghanaian environment. Trainers using this Field Guide outside Ghana should modify them to suit their training environment or better still source jokes that are relevant to the particular country.

**Jokes**

1. During the difficult days in Ghana, a driver wrote the inscription "Ghana is Hard!" on his vehicle. He was arrested and fined heavily. After that, he changed the inscription to "It is still Hard!" However, he was no longer put before court because the inscription did not mention Ghana or anybody's name.

2. During the economic slump in the early eighties, a man visited a lady. After a long conversation, he went to drive his car home. Lo and behold, all the four tyres under the car had been removed. A note on the windscreen of the car reads: *"If you love my girl friend, I also love your tyres".*

3. A long time ago, when university students were served at meals, supper was served late on a Sunday. The Church bell was rung and no student worshippers stepped into the chapel. The reverend minister went out and saw prospective student worshippers standing at the dining hall gate. In response to the call from the minister to come and worship, the students started demonstration with the chorus: "No Abintsi no Hallelujah" (Abintsi is a Hausa word meaning food and Hallelujah is a Church service singing, an exclamation of praise to God.)

4. Two rats went out to look for " abintsi". Unfortunately, both of them were caught in a trap after the older one warned the younger one that they had enough and had to return home. The younger one began to complain about the pains of the trap. The older one then explained that the pain they were going through was nothing serious and that they would experience the real pain when the trap owner arrived.

5. Long ago, a renowned scholar was invited to speak on the topic "The Floods". He prepared his notes very well; put on his best suit and as he was about to deliver his lecture confidently he raised his head to look around. Lo and behold, Noah himself was in the audience.

6. A professor was delivering an inaugural lecture in a university. The lecture was so long and boring that all the audience except one person left the hall. After the lecture, the professor thanked him for staying and enjoying the lecture. The man replied that he had no choice but to remain because he was an electrician and was only waiting to dismantle the expensive public address system being used.

7. On one Christmas Eve, two men informed their relatives that they would celebrate the Yuletide in style. Each of the guys went to the evening market for a good bargain. One promised to buy a turkey while the other promised to bring enough money from his sales for the celebration. In the market one bargained for a "big turkey" in a well-designed basket. There were not many problems in bargaining and both went home happily. The one who bought the "turkey", called the children to kill it for Christmas meals.

The other who sold the "turkey" called the children to come and share the money derived from sale of the turkey. When the children came to kill the "turkey", it turned out to be an old vulture.

To the dismay of the other family also, the money brought home were all fake currency notes totaling ¢100,000.

8. A pastor who frequently preached that there were no ghosts went to bury a corpse when it was getting dark. When he wanted to raise his hand to emphasize a point in the sermon, he realized that something was holding his gown to the ground. He started shouting and casting out the ghost from holding his gown. It turned out that it was his umbrella, which had pierced his gown as he fixed it on the ground to have elbowroom to preach the sermon.

9. An elderly man who was invited as a witness in an accident case was cross-examined by a lawyer. The elderly man stated that the distance between the victim and the vehicle after the accident was exactly 101 metres. The lawyer asked him why and how he came by the measurement. He replied that he took the measurement with his son's assistance, because he knew that some foolish man would one day ask him that question.

10. A gentleman took his girl friend to a restaurant for lunch. He placed order for fried rice and chicken. The illiterate lady thought of how to impress her lover about her intelligence. She then confidently called for smoked rice and a cock.

11. Togbe and Sowu went to a chop bar to eat. Togbe bought plenty of meat while Sowu could not afford to buy even one piece of meat. While they were eating Togbe regularly looked at Sowu's plate and smiled teasingly. When they finished the meal and were leaving the bar, Togbe asked Sowu what the time was, because the former did not own a watch. Sowu told him gently to ask his meat what the time was.

12. **Obituary Notice in a shop:** We deeply regret to announce the sudden death of Mr. Credit. The late Mr. Credit was survived by three children namely a Bounced Cheque, Unpaid Invoices and Unfulfilled Promises. Please take note and do not ask of him here. Thank you.

13. A couple lived unhappily because the husband was very difficult. He beat his wife severely anytime she misbehaved and complained or if the wife wanted to suggest something good to him. The husband and wife each had a dog. While the wife named her dog "You dare not say it"; the husband named his dog "Say it and you will receive a terrible blow on your mouth". So the couple lived thereafter without much verbal communication.

14. A couple was not on speaking terms and, therefore, always used the sign language to communicate. The husband was to travel by plane at dawn and therefore left a note for the wife to call him at 4:00 a.m. so that he could catch the plane at 5:00 a.m. At exactly 4:00 a.m. the wife wrote a note stating "It is exactly 4:00 a.m., please wake up and prepare for your travel".

    Unfortunately, the man woke up at 5:00 a.m. to see the note. He had missed a very important international meeting, which he was to chair.

15. An unfaithful wife and her boy friend had a bitter experience they would never forget. As both were in the room, the husband of the woman arrived. The boy managed to hide under the bed. The couple's young son then saw the young man's red face and exclaimed, "Look at the red face staring at me. I am waiting to see where you will sleep today". The son kept on saying this till his father wanted to find out the truth of what he was saying. Lo and behold, there was his close friend under the bed.

16. A wise old man became very popular in his community because he was able to solve all types of problems through advice to people far and near. A young man in the village was jealous about the wisdom and popularity of the old man and, therefore, planned to play a trick on the old man in order to disgrace him.

    He caught a young bird and went to the old man. His idea was that he would ask the old man whether the bird in his palm was alive or dead. If the old man should reply that the bird was alive, he would squeeze it to death. If on the other hand the old man stated that it was dead, he would let it fly to show that the old man was not a wise man.

    When he approached the old man with his question, the old man replied that the bird was in his hand and whether it was dead or alive was left to him to determine. The boy then went home sad after the old man has proved that he was really a wise man.

17. A teacher asked her students to demonstrate what happens in their homes daily. Some demonstrated how to brush their teeth and eat. A brother and his sister began to fight fiercely to the surprise of all. When they were separated and asked to explain why they were fighting, they replied that that was what their parents do daily.

18. Awo a debtor saw Esi her creditor approaching her house. She then told her daughter Afi to tell Esi that she had travelled. When Esi arrived and asked Afi her mother's whereabouts, the reply was that her mother had travelled. When Esi asked when the mother would return, Afi replied, "Let me go to the room and ask her when she would come back".

19. Two friends Asamoah and Asempa left their village to work in the urban centre. Their intention was to return home to surprise their relatives and friends with their achievements in the urban centre. Each of them, therefore, bought a big box to store wealth in it.
    Asamoah stocked his box with the tins of sardines and mackerel, which he consumed to show how he enjoyed life in the city. Asempa bought kente cloth, all types of dresses and clothing and gold necklaces. Asamoah realized that his box was not heavy enough to impress the villagers so he put stones in it. When both of them reached the village on their return, those who carried the boxes praised Asamoah for his hard work and wisdom. However, when the boxes were about to be opened, Asamoah was nowhere to be found.

20. Allah invited two men to make any requests which when granted would help them improve upon their lifestyles. The first person asked for education and knowledge that would help him enhance his performance. The other person happily noted that he had big barns of cereals and large stocks of tubers of yam and cassava and did not need anything.
    Over a period, Allah called them again to account for their stewardhsip. The first person enumerated the wonderful equipment

and tools such as electricity, computer, telephone, vehicles and aeroplanes that he has invented. The second person sadly noted that he has been following the first person, enjoying the facilities he has invented.

21. During a sermon, a Pastor began to advise the congregation against their involvement in certain vices. There was also a drunkard among the congregation who could not keep his mouth shut. Any time the Pastor mentioned any of the vices, the drunkard would retort and shout "Tell them". The sermon and the interruptions went on like this:

    Pastor:     According to God, thou shall not steal.
    Drunkard:   Tell them.
    Pastor:     Thou shall not gossip.
    Drunkard:   Tell them.
    Pastor:     Thou shall not defraud your friends.
    Drunkard:   Tell them.
    Pastor:     Thou shall not get drunk.
    Drunkard:   Who told you that? Have you ever been to heaven to speak to God?

22. A man who used to swim regularly with a domesticated crocodile defied all pieces of advice to stop swimming with it. One day, he returned from the sea with only one arm.

23. Participants in a literacy class were enjoying a radio programme in the local language when the batteries became weak. A young man was asked to rush and buy new batteries so that they could continue listening to the programme. When the new batteries were fixed, it was an English programme which was on air. The course prefect then assaulted the young man showering insults on him as to why he bought English speaking batteries.

24. When a diabetic patient went for a review at the hospital, the doctor was surprised that the patient's condition did not improve. The patient reported that he has been eating a lot of salted fish (*kobi*) since he was prevented from taking salt in his food.

25. A hypertensive patient was advised by his doctor to eat only one ball

of kenkey instead of six which the patient was used to for a meal. When the blood pressure level rose, he went to report at the hospital. In a reply to the doctor's question, the patient noted that, all the six balls were combined during the preparation, to make one ball of kenkey for him to eat.

## Proverbs

1. The beard does not make a person a philosopher.
2. When we reach the river, we shall know how to cross it.
3. You should not blame the rat alone: You have to blame the dawadawa also.
4. A hawk never goes on retirement.
5. If size ensures superiority, the elephant will come to town.
6. A man who does not lick his lips cannot blame the harmattan for drying them.
7. One does not drop a stick to fight with his bare hands.
8. One leans in the direction of his strength.
9. Success is the best perfume.
10. Knowledge is like a garden: if it is not cultivated it cannot be harvested.
11. A rat that wants a full stomach should not be afraid of getting wet.
12. When there is a hole in a tree, bees can produce honey in it.
13. If you wish to use a beautiful cloth to bury your grandmother you better buy the thread during her lifetime.
14. If you see a mother shedding tears it is not just because her child is hungry.
15. Hair that is infested with lice is not plaited.
16. You cannot farm along a footpath and say that no one should step on your farm.
17. A woman with a very broad back cannot carry two babies on it at the same time.
18. If you don't know where you are going, you must at least know where you are coming from.
19. If you don't know where you are going, even if you arrive there you will not realize it.
20. A patient dog eats the fattest bone.
21. A faint heart never wins a fair lady.
22. A child who cries persistently forces the mother to reveal what was eaten.

23. If you hold a snake by the head what is left is a rope.
24. However small the space, the hen will find a place to lay her eggs.
25. It is better to preserve one's respect even if one is poor.
26. Power is like an egg; when held too tightly, it cracks; when held loosely it falls.
27. When the roots of a plant are rotten the leaves will show.
28. A dog will not forsake its owner to follow a king or a rich man.
29. The tree which refuses to dance is forced by the wind to do so.
30. No one buys a cock only to let it crow in another person's house.
31. A slippery place does not know who a chief is.
32. If you are looking for something upstairs and you cannot get it, you better take it if you find it downstairs.
33. We do not put a live bee in a pot to make honey.
34. The legs of a hen do not kill its chicks.
35. A male bird has no nest.
36. A bald person has no business in the barber's workshop.
37. If you want to stop the smoke you have to put off the fire.
38. If you are not mad but pretend to be a mad, people will call you a mad person.
39. One tree cannot make a forest.
40. God does not give horns to a wicked animal.
41. A mouse does not dance to the tune of a cat's music.
42. No matter the height of a tree it cannot prevent the sun from shining.
43. If you want to cure a sick person, do not look into his eyes.
44. The young crocodile will never be killed in the presence of its mother.
45. An empty sack can never stand upright.
46. If you do not want to marry a quarrelsome woman you better marry a woman who has lost her teeth.
47. No matter how big an elephant is, it cannot use a cobra as its chewing stick.
48. Your enemy can never give you a good hair-cut
49. One does not know what he has missed until he has come across it.
50. A careless word may wreck a life but a loving word may heal and bless.
51. If you are pulling something and it cannot come down, you must know that something is hooking it: You must go and find out.
52. Fear is the greatest enemy of mankind.
53. Pepper is hot but worms live in it.

54. Water does not kill a crocodile.
55. Life is not a straight line: There are curves on it and some of the curves are very sharp.
56. When people are shaking hands, we know those who are left-handed.
57. If you do not ask for meat, you will continue to receive bones.
58. A single foot does not make a footpath.
59. Two eyes see clearer than one eye.
60. It is easy to know when war breaks out, but difficult to know when it will end.
61. The more you use power, the more you lose it.
62. No one takes appetizer for nothing.
63. It is easy to become a leader but it is difficult to remain a leader.
64. It is impossible for a poor man to live an independent life.
65. No matter how long the night, the day will surely come.
66. Even if you love a blind man you cannot give him one of your eyes.
67. Rudeness is a weak man's imitation of strength.
68. No one can lose what he has not got.
69. You cannot have a fair judgment of a chief's performance if you judge him by the way he sits.
70. He who reaps where he does not sow, will sow where he cannot reap.
71. Life is lived forwards but understood backwards.
72. Those who bring sunshine to the lives of others will never be denied the enjoyment of sunshine.
73. The wise man changes his mind at times but the foolish man hardly changes his mind.
74. If we all want to pay eye for an eye, the whole world will become blind.
75. The future belongs to those who have time to wait.
76. Positive change comes to those who struggle for it.
77. Those who wait to seek God in the 11th hour, die at 10:59.
78. When things go wrong, do not go along with them.
79. The dream of the dumb remains in his head.
80. You cannot burn the bridge and cross the river.
81. Anyone who speaks to you about others, speaks to others about you.
82. Any good thing done at the wrong time is sinful.
83. The horns cannot be too heavy for the animal that bears them.
84. Not all sins are crimes but all crimes are sins.

85. A tree which is not taller than you cannot give you shade.
86. Among the many sympathizers only a few are sincere.
87. If you are from the chief's house you need not go to meet drummers in the street.
88. Good deeds do not rattle.

## Some Thoughts for the Day

1. There are three classes of people:

   *Few who make things happen,*
   *Scores who watch them happen,*
   *Milliards who have no idea what is happening.*

2. It hurts to love someone and not be loved in return, but what is more painful is to love someone and never find the courage to let that person know how you feel.

3. A sad thing in life is when you meet someone who means a lot to you, only to find out in the end that it was never meant to be and you just have to let it go.

4. It takes only a minute to get a crush on someone, an hour to like someone, and a day to love someone but it takes a lifetime to forget someone.

5. Don't go for looks; they can deceive. Don't go for wealth; even that fades away. Go for someone who makes you smile because it only takes a smile to make a dark day bright.

6. The happiest people do not necessarily have the best of everything, they just make the best of everything that comes along their way.

7. Love begins with a smile, grows with a kiss and ends with tears.

8. When you were born, you were crying and everyone around you was smiling. Live a commendable life so that when you die, you are the, one smiling and everyone around you is crying.

9. Teaching a pig to sing may be very difficult indeed but it can also be most annoying to the pig. (Preware Youdeowei)

10. If you have never been loved, you will not appreciate the meaning of affection and, therefore, you can never give love. (Anthony Youdeowei)

11. Education for children is like filling a cup with tea, milk and plenty of sugar while adult education is more like stirring an already full cup of tea to blend the ingredients in a new way.

12. I have no shoes and complained, until I saw a man who has no feet.

13. The Good Lord gave us two ends, one to sit and another to think; the level of our success in life depends on the extent to which we use both ends.

14. Almighty God gave us two ears to listen extensively and one mouth to speak as little as possible and only when necessary.

**Reflections**

1. **One cannot grow without taking some risks**

**The Dilemma (Ann Sanders)**

To laugh is to risk appearing a fool.
To weep is to risk appearing sentimental.
To reach out for another is to risk involvement.
To expose feelings is to risk rejection.
To place your dreams before the crowd is to risk ridicule.
To love is to risk not being loved in return.
To go forward in the face of overwhelming odds is to risk failure.
But risks must be taken because the greatest hazard in life is to risk nothing.
The person who risks nothing, does nothing, has nothing, is nothing.
He may avoid suffering and sorrow.
But he cannot learn, feel, change, grow or love.
Chained by his certitudes, he is a slave.
He has forfeited his freedom.
Only a person who takes risks is free.

## 2. The struggle of the African people

It is a struggle of the African people inspired by their own suffering and their own experience. It is a struggle for the right to live.

During my life I have dedicated myself to this struggle of the African people. I have fought against white domination, and I have cherished the idea of a democratic and free society in which all persons live together in harmony and with equal opportunities. It is an idea that I hope to live for and to achieve. But if needs be, it is an ideal for which I am prepared to die.

Nelson Mandela
20/4/64

## 3. Procrastination is the Thief of Time (Edward Young, 1742)

Stop! Don't put this article down! You know what might happen. You might put it down and say: "That's an interesting title, but I don't have time to read it now. I'll get to it later." But later may never come.

If we can and should take action now, delaying needed action could cause more problems later, and then delaying is procrastination. For example, washing dishes after they've sat for some time makes it harder to scrub them clean. Postponing car maintenance can result in costly repairs later. Falling behind in paying a bill can result in heavier charges or the loss of service. One woman calculated that her overdue traffic tickets, videotapes, and library books totaled 46 dollars in late fees! That was for just one month!

Different people procrastinate at different stages. Some procrastinate before starting because they view the project as too big. Others begin, but about halfway through, enthusiasm wanes, and they put off finishing it. Still others get close to completing it but start another project, leaving the first unfinished. (You're doing fine, by the way. You're already halfway through this article).

Your reasons for not starting or completing a project may fall into all three categories. In the book the *New Habit,* Neil Fiore wrote: "The three main issues that are at the bottom of most procrastination problems are: feeling like a victim, being overwhelmed, and fear of failure." Whatever the reasons, if you can put your finger on the causes, you'll be closer to the solution.

If you are uncertain why do you procrastinate? Make a log of your activities for a week by half-hour intervals. Determine how you're spending time. It can be a real eye-opener to see how much time we spend on relatively unimportant things between important tasks. But then what? Think of the consequences.

Expecting that something will get done without putting effort into it can produce a sickening feeling. As you get closer to the expected deadline, you begin to feel pressure and anxiety. As these feelings build, your creative ability may be hampered. You are not as inclined to measure or weigh various ways to accomplish the goal but are mainly interested in getting it done.

For example: You're assigned to give a presentation. The night before, you sit down to get a few words on paper. You have not spent enough time to research your subject so you wing it. Perhaps with just a little more effort, you could have included experiences, supporting information, or charts to help your audience visualize the subject.

Another consequence that comes when we delay a project is the inability to relax when we have free time. That's because we have a nagging feeling (or a nagging someone who reminds us) that we have left a project undone.

**Wise Sayings Associated with Great Men**

He who walks with the wise grows wise (*Proverbs* **13.20a**).

As a well-spent day brings happy sleep, so life well used brings happy death (Leonardo Da Vinci, 1432–1519).

The mind of the guilty is full of scorpions (William Shakespeare).

Live your life and forget your age (Frank Bering, 1877–1965).

Success is a habit (Aristotle).

I am not interested in the past. I am interested in the future, for that is where I expect to spend the rest of my life (Charles Kettering, 1876–1958).

A man's life is what his thoughts make of it. (Marcus Aurelius, 211–180).

The excess of any pleasure degenerates into evil (Rousseau)

There are three faithful friends — an old wife, an old dog, and ready money (Benjamin Franklin, 1706–1790).

It is better to have loved and lost than never to have loved at all (Alfred the Great).

The unexamined life is not worth living (Aristotle).

All they that take the sword shall perish with the sword. (Matthew 26: 32).

Man is born free but every where he is in chains (Thomas Hobbes).

Who then is free? The wise man who can govern himself (Horace, 65–8 BC).

I found Rome a city of bricks and left it a city of marbles (Augustus Caesar, 63AD).

There is nothing permanent except change (Heraclitus).

All the resources we need are in the mind (Theodore Roosevelt, 1858–1919).

We will either find a way or make one (Hannibal).

A cheerful look makes a dish a feast (Aurelius Clemens Prudentius).

A man of words but not deeds is like a garden full of weeds (Anonymous).

## Ten Important Guides to Fulfilling Life

1. The most endangered species in Africa–dedicated and visionary leaders.
2. The most powerful force in life — Love
3. The most powerful attire — Smile
4. The most important asset in life — God
5. The most prized possession — Intergrity
6. The most valued property — Hope
7. The most destructive habit — Worry
8. The deadliest weapon — Tongue
9. The greatest enemy of mankind — Fear

# RECOMMENDED READING

Amedzro, Albert (2005) *Globalization, Non-Formal Education and Rural Development in Developing Economies.* Accra: Ghana Universities Press.

Beardwell, I. and Holden, L. (1994) *Human Resource Management: A Contemporary Perspective.* London: Pitman Publishing.

Bown, Lalage and Tomori, S. H. Olu (1979) *A Handbook of Adult Education for West Africa.* London: Hutchinson University Library for Africa.

Dorr, Donal (1990) *Integrated Spirituality: Resources for Community, Justice, Peace and the Earth.* Harare: Gill and Macmillan.

Fanon, Franz (1967) *The Wretched of the Earth.* London: Penguin Books.

Finlay, Mary (1994) *Communication at Work.* Toronto: Harcourt Brace and Company.

Freire, Paulo (1970) *Pedagogy of the Oppressed.* New York: Seabury Press.

Hope, Anne and Timmel, Sally (1984) T*raining for Transformation: A Handbook for Community Workers,* Books 1, 2 & 3. Gwen, Zimbabwe: Mambo Press.

Knowles, Malcolm (1970) *The Modern Practice of Adult Education.* New York: Association Press.

Kunczik, Michael (1984) *Communication and Social Change.* Bonn: Friedrich-Ebert Foundation.

Luft, Joseph, (1970) *Group Process: As Introduction to Group Dynamics.* New York: Mayfield Publishing Co.

Maslow, Abraham H. (1970) *Motivation and Personality.* New York: Harper and Row.

McGregor, Douglas (1960) *The Human Side of Enterprise.* New York: McGraw-Hill.

Mullins, L. J. (1993) *Management and Organizational Behaviour.* London: Pitmans.

Pretty, J. N., Guijt, I., Thompson and Scoones, J. I. (1995) *Participatory Learning and Action: A Trainers Guide.* London: IIED.

Srinivasan, Lyra (1973) Non-formal adult learning. World Education, New York: USA.

Stanley, John (1982) People in Development. *Search,* Bangalore.

WARDA (1991) *Training of Agricultural Trainers: Course Instructor's Manual and Course Handouts.* Bouake, Cote d'Ivoire: West Africa Rice Development Association.

Youdeowei, Anthony (2001) *Principles of Integrated Pest Management. Growing Healthy Crops*. IPM Extension Guide Number 1 PPRSD/GTZ/ICP Ghana.

www.ingramcontent.com/pod-product-compliance
Lightning Source LLC
Chambersburg PA
CBHW061417300426
44114CB00015B/1974